1995

Research *and* Data Analysis

in

LEISURE, RECREATION, TOURISM *and* SPORT MANAGEMENT

Joe M. Shockley, Jr.

Sigma Press
Las Vegas

Research and Data Analysis in Leisure, Recreation, Tourism and Sport Management

Library of Congress Catalog Card Number 95-69634

ISBN Number 1-57087-164-7

Published by Professional Press
Chapel Hill, North Carolina 27515-4371

Manufactured in the United States of America
96 95 94 93 92   10 9 8 7 6 5 4 3 2 1

Dedicated to the memory of my beloved wife, *Marilyn Ropp Shockley*
for whom no adequate positive statistic has ever been defined.

Preface

This text is written for a first course in research and data analysis in leisure, recreation, sport management and tourism. There are many good generic texts that address the subject adequately, but over the course of using these texts for the last 20 years, the author has observed students with difficulty relating generic concepts to their field. By using examples from leisure, recreation, sport management and tourism, the author has found students more interested in and capable of mastering introductory research and data analysis concepts.

A trend has been to send leisure, recreation, sport management and tourism students to other departments for research and data analysis classes. However, there are many qualified instructors in the field who can relate professional experiences better than examples of rats running a maze, numbers of students who can memorize nonsense syllables, or find dollar variance in two columns of numbers. It is only when the field itself assumes more responsibility for data collection and analysis that students will enjoy becoming better administrators, practitioners and researchers.

In addition to using examples from the field, the methods used are more from a practical than a theoretical base. For example, symbols for procedures, such as sums of deviations squared (Σx^2, SS) are often used interchangeably in many texts without making clear what the symbols mean. A student who is looking for a formula approach may see several variations of correlation procedures published and not realize they are the same methods reduced to lowest terms in one case, while including all unaltered terms in a different case.

This text uses a method that gives insight to relationships in different statistical procedures. By beginning with a basic concept, the sum of deviations squared, connections between variance, standard deviation, correlation, t and analysis of variance are more easily understood. In addition, finding total, between and within sums of squares *the same way* in both one-way and multiple analysis of variance, associations in both procedures are more apparent to the beginning student.

Part 1 introduces research methods and report writing, emphasizing descriptive research and experimental design.

Part 1 introduces research methods and report writing, emphasizing descriptive research and experimental design.

Part 2 uses **grouping data** as an introduction to graphics and data analysis. Grouping data is helpful when large amounts of information needs to be displayed clearly on one page. Grouping data is also a key to understanding graphics.

Part 3 examines elementary statistical concepts using **raw scores** including central tendency, variability, hypothesis testing, sampling and both parametric and non-parametric correlation. These techniques are useful for descriptive and correlational research. If graduate students have already been introduced to grouping data, they can begin with Chapters 1-7, skip to 10 and easily master the rest of the text in a single semester.

Part 4 introduces concepts of **group associations** including t testing, one-way analysis of variance, multiple ANOVA, non-parametric ANOVA and chi square. These associations are useful in experimental design research.

The approach used in this text with both grouped and raw score data lends itself well to computer spreadsheet techniques. By typing data in columns and rows, and inserting cell formulas at convenient places in the worksheet, much of the tedium and many minor math errors can be eliminated. Also, the graphics discussions make transition to major spreadsheet graphics much easier than trial and error on a computer.

The author is indebted to the many **students and teachers** who taught him the importance of data analysis, and to the copyright holders who generously allowed the reproduction of selected data and tables, without which the text would not have been possible.

More specifically, the author is grateful to the **Longman Group, Ltd.**, on behalf of the Literary Executor of the late Sir Ronald A Fisher, F.R.S., and Dr. Frank Yates, F.R.S. for permission to reproduce Tables III, IV, VII from <u>STATISTICAL TABLES FOR BIOLOGICAL AGRICULTURAL AND MEDICAL RESEARCH 6/e (1974)</u>.

Also, **The American Statistical Association** and M.G. Kendall and B. Babington Smith for a table of random numbers; **The Biometric Society** and Leon H. Hunter for a table of critical values for Duncan's New Multiple Range Analysis; **The Institute of Mathematical Statistics** and E. G. Olds for a table of significance for rank differences; **The Iowa State University Press** and G.W. Snedecor, for a Table of F; **McGraw-Hill, Inc.** and Sidney Siegel for a table listing nonparametric statistics; **The Psychological Bulletin** for reproduction of a normal curve; **The Research Quarterly for Exercise and Sport** and Charles C. Cowell for a Personal Distance Inventory and **Houghton Mifflin Company**, Donald T. Campbell and Julian C. Stanley for permission to modify work from Experimental and Quasi-Experimental Designs for Research.

Contents

Part 2 Grouped Data

Part 3 Raw Scores
Central Tendency and Variability

Symbols

<	Less than.	>	Greater than.
s, σ	Standard deviation.	≠	Not equal to.
σ	sigma; population standard deviation.	Σ	Sigma. Sum of.
≥	Equal to or greater than.	≤	Equal to or less than.
μ	Mu; population mean.	$\bar{x}$	Mean.
X²	Score Squared.	**s²**	Variance.
x	Deviation score; usually **X** - $\bar{x}$	**x²**	Deviation squared.
Σ**x²**	Sum of deviations, squared.	Σ**X**	Sum of scores.
Σ**X²**	Sum of scores, squared.	**X (Y)**	Score.
α	Alpha.	β	Beta.
P	Rho.	χ	Chi.
|**N**|	Absolute value of the number N.		

$P_{(N)}$ Specific percentile, defined by N. Score at or below which a given % (N) lie.

$PR_{(N)}$ Percentile rank. Number (%) representing a group of scores that ≤ the given score (N).

$C.I._{(.05, .01)}$ Confidence interval (95%, 99% level, respectively).

Part 1

The Research Process

Chapter 1

What is Research?

The media today are full of research statistics. Headlines often exclaim, don't eat this type of food - it results in high cholesterol. Or one popular book tells us that research has shown if we exercise aerobically, it will keep us trim, while a different source says a group of people who walk 20 minute miles burn more fat than those who jogged 10 minute miles. The results of presidential elections are known long before the final vote is tallied. With results reported that are often conflicting, yet often strikingly accurate, just what is this word called research?

Merriam (1949) defined research as:

1. Careful search; a close searching. 2. Studious inquiry; usually, critical and exhaustive investigation or experimentation having for it's aim the revision of accepted conclusions, in the light of newly discovered facts.

By the above definition, research has something to do with critical and exhaustive investigation, and it may bring to light new conclusions. Yet the media are full of conflicting and contradicting judgements, often about the same subjects. Why doesn't all research result in consistent findings?

Part of the answer lies in the use of the word *exhaustive*. How many studies are truly exhaustive? Often, a study results in a cursory review of literature, isolates and focuses on specific measures or factors, violates assumptions in data analysis and returns results that are sometimes incomplete, incorrect and inconclusive. Is there any wonder why inconsistencies sometimes occur?

Yet in spite of inconsistencies, students and practitioners alike are called upon to do research. Business firms spend a lot of money on it; the quality of life we enjoy today evolved from it. Just what kinds of research are used in the sport management, recreation and tourism fields?

4 What is Research?

1. Types of Research.

Another piece of the puzzle may lie in research applications. Steinhaus (1965) described research as *basic* and *applied*. Basic research, according to Steinhaus, was like a thermometer; it is a technique designed to measure, record and report results under existing conditions. Applied research was more like a thermostat; it relies on accurate measurement, but it does something useful with the conclusions. See Figure 1.1.

Basic:	Applied:
Theoritical constructs.	Immediate answers.
Uses animal subjects.	Uses human subjects.
Carefully controlled.	Less controls.
Results may lack application.	Results useful - practical.

Figure 1.1. Basic Versus Applied Research.

All research is oriented toward *prediction*. Thus, basic research may uncover basic truths that can be used to predict results under similar circumstances. By the same token, applied research predicts results, but the prediction may not be as accurate or consistent as basic research.

Another distinction between basic and applied research may be that basic research is oriented more towards discovering *principles* - facts that remain the same anywhere so long as conditions remain the same. Applied research may also discover principles, but is oriented more towards practical results that can be better explained by *theories* - a more or less plausible scientifically accepted explanation that seems to define the facts associated with the phenomena, but which may need to be altered occasionally by new evidence.

So when we are bombarded by research that produces inconsistent conclusions, distinguishing between that which is basic and that which is applied may lead to a better understanding of why the results are not always similar. Authority figures may inadvertently discourage research; it is difficult to prevail against modern science. But if research is to yield results that are predictible, it should follow the scientific method.

2. The Scientific Method.

ust what is the scientific method? Aristotle once ruled the scientific world for several hundred years with his analysis that proved the earth was the center of the universe. Today, there are many scientific procedures that help eliminate this type of error, but the major way is through a systematic, logical approach. It involves at least 4 distinct steps. They are:

A. define and limit the problem. As strange as it sounds, a problem is not always easy to define. It requires being very specific as to what is meant by the problem description. And if the problem area is too broad, focus may not be possible. In most cases, limit the study to a few variables at most. Some ways to recognize a problem are:

deviation from some standard. Have you ever noticed that something just didn't fit the usual standard? Clyde Tombaugh noticed the planet Neptune wasn't exactly where it should be and predicted an unknown body would be found close by - it turned out to be Pluto. Whenever something seems to be different from what it should be, there is a possible research project.

viewpoint. What do other people say about a possible problem? Often a thesis or dissertation will contain a last section called implications for further study that define possible research.

empirical. Empirical data is information that comes to us through our senses. Often it is incorrect, but sometimes it is worth investigating. Archimedes noticed water overflowing in his bath and used it to discover the principle that a body displaces it's weight in water. Observation of changes in interaction by students in a games class led the author to a sociometric study of social distance.

literature. The literature is full of ideas for further research. A study done at one level, such as high school, may be appropriate for a related study at a different level, such as college. A consistent finding of results of a cold hip bath in a major professional journal over a number of years led to a study using cold water as an ergogenic aid for cross country runners (Shockley, 1977).

B.   hypothesize. A hypothesis is an expected result. If the direction is not known, the hypothesis can be stated in the null (no difference) form.

example: There will be no difference between female and male preference for a training film on safety.

C.   gather data. The use of instruments to gather and record data is crucial. Instruments can be electrical, mechanical, written, etc. Instrument objectivity and reliability is an important consideration. The kind of controls used to gather the data is also important. Two kinds of validity (does the instrument measure what it is supposed to measure) are:

internal validity. Results need to be attributed to different treatments; therefore, all variables should be controlled.

external validity. External validity refers to how general the results can be applied. Sampling procedures need to be scientifically controlled in order to generalize to the total population.

D.   analyze and interpret results. Often, the results are quantified with numbers, so statistical analysis is appropriate. It is important that the statistical procedure that is used meet the data assumptions in order for the analysis to be meaningful. Generally, the most powerful technique that meets the data assumptions is suitable.

3. Challenges to Scientific Inquiry.

Today, challenges are being made to the normal scientific method by those who claim all research is qualitative and relative. Schiapirelli once found canals on the planet Mars. We now know these observations were in error. Modern physics uses ideas from Lorentz and Einstein that show measurement of an event depends on what frame of reference the observer is in. Depending on the frame of reference, the measurement may not be as quantifiable or as accurate as was once thought.

For example, time measured by a clock on planet Earth observing a space vehicle traveling near the speed of light is not the same as time measured within the vehicle itself. Timing two points in space with two separate clocks would result in two different numbers. Both measurements would appear to be absolute - and they would be - but only within their own frame of reference. In this regard, there is no absolute time. Using a similar analogy, how can we be so sure of measurements taken on our own spinning, wobbling planet?

Also, the very act of observing a phenomenon may alter it's behavior. Again, modern physics has found that in trying to measure the innermost part of an atom - the quark - the very conditions required to measure it affects it's behavior and results in a different number than when no attempt is made to measure it. So what does the measured number really amount to? The meaning is that the very building blocks of matter cannot be reduced to *absolute* building blocks with known mass or electric charge, although the last known quark has just been found.

Yet, at the turn of the century, Lorentz found that certain characteristics of light did not have the values predicted theoretically, so he concluded there was an ether of unknown quantity. Other scientists agreed that either there was an ether or an interferometer arm had to shorten very slightly in order to account for the difference. Again, observing - measuring an inanimate physical object - changed the very properties physical scientists were trying to quantify.

So there are challenges to scientific inquiry. Perhaps Schiapirelli wanted those Martian canals to exist. No one has accused him of deliberately altering what he saw - his eyesight simply betrayed him. But we are all human and subject to the Schiapirelli effect, and what we measure may be only relative to our small world. Does that mean that research is really not quantifiable, and does it follow that research is of little value?

Certainly not! The measures we have may be gross by an absolute standard, but they are accurate enough to give us laser telecommunications, desktop computers and space ships that show some human progress, even if it is gross. The research process has produced many modern benefits, and must be continued, albeit with a slight sense of skepticism.

4. Kinds of Research.

Although this text will focus on descriptive and experimental research methods, an overview of these and other kinds of research is presented here. An excellent categorization of research methods is found in Thomas and Nelson (1990). These methods include analytical, experimental, descriptive and qualitative. The descriptions given are not intended as inclusive; rather, they are listed as a gross introduction for concept development only. Qualitative research will not be discussed.

Analytical

A.   Historical. Historical research includes events that are analyzed according to facts that were accumulated from the past. People are often researched, as are trends, events and other happenings. Whatever the focus, the procedure seems to follow certain steps, all of which may not be necessary, and not always in the order presented:

1.   gather data. *Primary* data are observed first hand and written down. Only one mind exists between the data and the recording of it. Primary data may include documents recorded by others. Relics and artifacts are also primary, but the researcher has to interpret their meaning.

Secondary data are already recorded by others. Another mind has already recorded and interpreted the information, and at least one additional mind, the researcher, reads or examines the data before interpreting it further. Primary data are preferable to secondary sources.

2.   subject the data to criticism. *External* criticism seeks to discover the authenticity of the data (Van Dalen, 1965). For example, when the Shroud of Turin was subjected to Carbon 14 dating, it was found the material came from a time several hundred years A.D.

        Internal criticism seeks to determine the validity of the source. For example, the words and patterns of the Dead Sea Scrolls were found to be almost identical with later versions of the Bible, showing a consistency that was explained by some scholars as internal validity.

3.   state the problem. After gathering data and criticizing it, the nature of the problem under investigation becomes more apparent.

4.   hypothesize. At this time, hypotheses or guiding questions may or may not be formalized.

5.   results and conclusions. Return to the data obtained to support or reject hypotheses, and report the results and conclusions of the study.

B.   Philosophical. A philosophical study uses *induction* or gathering facts that can be synthesized into a general principle or theory to explain a phenomenon; or *deduction* - generalizing from a principle or theory to explain particular facts that describe a phenomenon. As such, it is a scientific way to research not only what is, but also what should be. There are no specific steps; a study may develop a standard and judge facts by the standard, or in explaining what exists, a taxonomy or classification system may be used. Some of the procedures a philosophical study may encompass are (Metheny, 1965):

    1.   identify assumptions relating to a problem.

    2.   define the problem.

3.   gather and interpret facts.

4.   formulate hypotheses.

5.   design the study - gather facts.

6.   analyze the facts.

7.   discuss the findings and conclusions.

C.   Meta Analysis. In research reported in the literature, many studies use different numbers of subjects or use different techniques that make comparisons of similar studies difficult. By using mathematical treatment of means and standard deviations, different studies may be compared. Steps in meta-analysis are (Thomas and Nelson, 1990):

1.   identification of a problem.

2.   literature search.

3.   review the literature searched to find studies to include.

4.   identify and code the important study characteristics.

5.   calculate effect size.

6.   use further statistical tools.

7.   report steps and findings/conclusions.

Descriptive

A.   Questionnaire. In questionnaire research, the instrument used is a vital part of the process, as is selection of the sample. Steps may include:

1. define and limit the problem or topic to be researched.

2. determine assumptions and limitations of the study.

3. review the literature for similar studies.

4. develop or use a questionnaire. In order for a questionnaire to produce data useful for prediction, it should contain four elements:

    a) *validity* refers to whether the questions measure what they are supposed to measure.

    b) *reliability* means the questionnaire gives consistent answers.

    c) *objectivity* is consistent data obtained by different persons who administer the instrument.

    d) *freedom from group bias* occurs when consistent measurements are obtained from different groups.

5. use the instrument to gather data.

6. report results and conclusions.

B. Interview. Interview research is done basically the same as a questionnaire study. Procedures for the interview are usually written down and read to the person being interviewed. The interviewer needs to be carefully trained not to influence biased answers while recording data.

C. Normative survey. This type of research refers to establishing norms for data. For example, a maintenance supervisor may develop or use norms for time to repair a pothole, sweep a gym floor, etc.

Whatever the measures being developed as norms, a questionnaire or test is used to collect the data. Sampling techniques are vital; if the data is to be generalized to a population the sample must adequately represent the population. Percentiles are then computed which become the norms. Norms may be used as ranges rather than exact numbers.

D. Case study. A case study may use a very small number of clients, and as such the results are not often generalizable to a large number of people. However, an in-depth study of one or more subjects can produce useful information about those studied. Steps are (Rarick, 1965):

1. determine the value of the study.

2. obtain data.

3. analyze the data.

4. make recommendations.

5. appraise effectiveness when the recommendations are implemented.

E. Developmental. This type of study is one that is done over a period of time. It is much like a questionnaire study, with the major difference being the length of time involved. A special category of developmental research is *cross-sectional*, where different clients represent time, rather than using the same subjects over time.

F. Correlational. A correlational study examines a statistical relationship without determining cause and effect. Even though the relationship does not show cause and effect, the data may still be very useful. The reader will recall that initial warnings about smoking cigarettes were at first correlational. Only later did experimental design reveal the harmful cause and effect.

Experimental

In the experimental method, different variables are exposed to different treatments. A problem is that not all variables and not all treatments can be controlled. But through the use of randomization and other procedures, cause and effect relationships may be determined.

A *dependent* variable is one that is being produced - the yield. An *independent* variable is generally referred to as the treatment or experimental variable. Steps generally include:

1. define the problem.

2. determine assumptions/limitations.

3. review the literature.

4. apply the different treatments.

5. gather data. Data gathering may start prior to the experiment.

6. analyze the data.

7. determine the results and conclusions.

EXERCISES

1. Using the steps in the scientific method as a standard, compare how the philosophical, questionnaire, case study and experimental methods of research measure up to the scientific method.

EXAMPLE: Historical.

Standard	Met Criterion		How? or Why Not?
	Yes	No	
A. Define Problem	X		Stated after facts found.
B. Hypothesize	X		Often.
C. Gather Data	X		Primary & secondary.
D. Report Results/ Conclusions	X		Final step.

2. Explain the difference between applied and basic research. Which one is best? Why?

3. Why do similar studies often yield different conclusions?

4. Define the following terms:

theory	principle	subject
research	validity	empirical data
reliability	objectivity	
group bias	primary data	
secondary data	internal consistency	
external consistency	correlational study	
hypothesis	assumptions	
meta analysis	induction	
deduction	taxonomy	

References

Bethel, John P. (1949). (Gen. ed.) Webster's new collegiate dictionary (p.720). Springfield, Massachusetts: G. & C. Merriam Company.

Metheny, Eleanor. (1965). Philosophical Methods. In M. Gladys Scott (Ed.), Research methods in health, physical education, recreation (pp. 482-496). Washington: American Association for Health, Physical Education, and Recreation.

Rarick, Lawrence. (1965). The case study. In M. Gladys Scott (Ed.), Research methods in health, physical education, recreation. Washington: American Association for Health, Physical Education, and Recreation.

Shockley, Joe M. Jr. (September, 1977). Warmup with a cold shower? Track Technique. 2195-2197.

Steinhaus, Arthur H. (1965). Why this research? In M. Gladys Scott (Ed.), Research methods in health, physical education, recreation. Washington: American Association for Health, Physical Education, and Recreation.

Thomas, Jerry R. & Nelson, Jack K. (1990). Research methods in physical activity (pp. 19, 248). Chaimpaign, IL: Human Kinetics Books.

Van Dalen, D.B. (1965). The Historical Method. In M. Gladys Scott (Ed.), Research methods in health, physical education, recreation. Washington: American Association for Health, Physical Education, and Recreation.

Chapter 2

Whatever type of research is contemplated, the scientific method is used. This chapter will review this process, including selecting a problem, limiting the problem, developing hypotheses or guiding questions, literature search, collecting data, analyzing data and reporting results and conclusions.

1. Selecting a Problem

Just how does a person discover a problem that needs to be researched? Surely all the known problems that a beginner might explore have already been discovered by those already in the field. The beginning researcher is overwhelmed by the process, because time and other constraints demand the topic begin before a full understanding of the method has been developed. Further, persons of renown who are considered as authorities are thought of as modern Aristotles by the novice. Their knowledge cannot be challenged, and the beginning researcher may feel intimidated beyond all hope.

I remember when some of the best scientific minds - physicians - told a patient who had just experienced a heart attack to essentially rest for the rest of their lives. Yet a Physical Director in the Cleveland YMCA had a small group of cardiac patients do exactly the opposite, and today most hospitals start a person moving soon after this unfortunate experience.

Look for a moment at what some outstanding people have said. "Sensible and responsible women do not want to vote," (Grover Cleveland, 1905). "Heavier than air flying machines are impossible," (Lord Kelvin, President, Royal Society, 1895). "There is no likelihood that man can ever tap the power of the atom," (Robert Millikan, Nobel Prize in Physics, 1923). "Ruth made a great mistake when he gave up pitching," (Tris Speaker, 1921). "Who the hell wants to hear actors talk?" (Harry M. Warner, Warner Brothers Pictures, 1927). "Everything that can be invented has been invented," (Charles H. Duell, Director of the U.S. Patent Office, 1899).

Fortunately, educators have plodded on in spite of authority and difficulty (Carter, 1966). Pauling, a chemist, researched blood clotting and discovered the effect of prothrombin in the blood. Clyde Tombaugh must have felt great wonder at why no one else took the time to explore a known aberration in the planet Neptune - but it was he who discovered Pluto. Arthur Gates paid 966 subjects to either memorize or generalize nonsense syllables in spelling, and he learned memorization is better short term, but some scholars still ridicule memorization as a learning technique.

Even though Ptolmey placed Earth as the center of the universe, Copernicus exhorted the heliocentric theory and was banned from the church. As early as 600 B.C., Thales wrote about static electricity. Democritus, another of those ancient Greeks, recited the atomistic theory to explain the structure of matter. Although there was no known way to determine the difference between pure gold and fake gold, Archimedes found a way, saving his head and establishing the principle of buoyancy.

Napier developed logarithms in **1614**. John Dewey established what was probably the first laboratory school at the University of Chicago in 1896. W.E. Dubois saw the need for equal opportunity, and his persuasive movement led to development of the NAACP in 1909. B.R. Buckingham started a research movement that had as an outgrowth the *Journal of Educational Research*. In 1758, Ben Franklin wondered why England had a marine climate when it had the same latitude as Newfoundland. Two centuries later, in 1958, the gulf stream was identified. Ben Franklin liked to swim during his leisure time, but he was unable to swim against ocean current, so he invented swim fins. And what about the ancient Chinese who liked fireworks during their leisure time? At first, skyrockets were seen only as playthings. Then, someone got the idea rockets might deliver mail, but they lacked accuracy. So the play concept essentially remained until World War II, when German V_2 rockets began at White Sands, N.M. what is today the most challenging scientific inquiry of all time - the exploration of space. What if all these people and events had listened to authority or felt intimidated?

Most students start work in Sport Management, Recreation or Tourism without any idea of what a topic for a thesis or research project might be. It is too easy to grasp the first suggestion by an advisor or professor, or to pursue a vague outline based on experience or imagination, only to discover later that the topic is not interesting and not tenable. To be successful, a topic needs to be well chosen and carefully planned, and a **lot** of background reading is necessary.

Locating a Problem. Bookwalter (1965) listed 3 ways to help start the process. They were experience in a vocation, education and training, and a direct search.

Experience in a vocation can certainly be a help. A student who has a background in sports activities may have knowledge and interest to research a sports topic. While a city recreation director, Dave Dugan, who had considerable experience in swimming, completed a dissertation on the effect of training methods on age group swimmers while at the University of Georgia. Alton Little, a former recreation director who had experience in gathering data through questionnaires completed a descriptive research project using similar techniques. James Colbert had experience in exercise classes that led him to study the difference between walking and jogging as an aid to aerobic conditioning. A student in Vermont who had experience as a Little League coach and administrator used her knowledge to investigate "tennis elbow" in little league pitchers. Frank Torres, a student in tourism at New Mexico Highlands University used his knowledge and interest in the Las Vegas area to locate on an old aerial photograph found at the Soil Conservation Service where ruts from the old Santa Fe Trail came into town. Students have a variety of work experiences from which to draw in order to investigate a problem with which they are familiar.

Familiarity in any area of expertise may spark ideas. Tourism - what interests you about that? Demographic variables of clients? Points of origins of tourists? How many days the average tourist spends at a site? How much money a tourist spends for food, entertainment, lodging, transportation? How a tourist found out about an attraction? Modes of transportation? How effective a brochure or marketing plan has been? Room occupancy in local motels? Destination research? Fitness interests of tourists? Culture variations in travel?

Leisure services or recreation - any interest or experience here? Demographic characteristics of clients? Number of clients in special programs? Community needs analysis? Needs of special populations? Trends in activities? Biomechanical analysis of dance routines? Day or resident camp programs? Time/cost required for maintenance? Accountability? Health/wellness benefits of activity? Costs/benefits of wellness programs? Effectiveness of leadership styles? Attitudes towards disabled people? Effects of aerobic conditioning on depression? Perceptions of people with disabilities? The effect of drama on socialization? Agricultural pollution of wetlands? Cultural diversity in programming? Effects of off-highway vehicles on land management? Fitness assessment in older adults? Program evaluation? Teens and substance abuse? Juvenile delinquency? Graffiti prevention? Curbing vandalism? Current child-care practices? Departmental image assessment? Therapeutic effects of outdoor adventure programs? Ethics in leisure? Leisure behaviors? Playground safety? Resource deterioration? The point is there are many varied possibilities for research, depending on the background that a student has.

Education and training offer similar opportunities. A course in research offers basic introduction to research skills that will help locate a problem area. A study that includes measurement, experimental procedures, statistical models, etc. will require additional education in these fields. Training in corollary fields, such as business, philosophy, psychology, sociology, physiology, etc. can provide additional ideas for possible research. All formal education exposes a person to concepts and authorities who can provide ideas for the research project. Be aggressive, seek out these sources!

A direct search for a topic involves going to people or written records in order to obtain ideas. Ask teachers and practitioners what recommendations they have. Often a person will have an idea that they don't have time to explore but are willing to share. Examine theses, research articles and dissertations - concluding remarks often include ideas for further research. Attending workshops and professional conferences may lead to a creative thought. Carry note cards so that any idea that presents itself while listening to someone else's presentation can be jotted down. Going to the library and electronically or manually scanning the literature is likely to be a waste of time, unless the researcher is already on the track of an idea.

2. Limiting the Problem

Once a possible topic is located, it will be necessary to limit the scope of the study. A broad topic needs to have focus-there just isn't time to explore every aspect of the research. Furthermore, a broad topic may result in information overload that misses the mark. For example, a person familiar with sports programs may decide on the effects of sports as a topic to investigate. But at what *level?* Elementary, middle, high school age? Adults? Then, perhaps what *sex?* A further delimitation may be what *type* of sports? Individual? Team? Baseball, basketball, football, softball, volleyball, etc.? So a revised topic now becomes "the effects of recreation baseball on middle school boys." Now there may be enough focus to begin reading about this topic.

A quick look through an electronic search such as InfoTrac could result in discovering that there are actually many kinds of possible effects - scholarship, injuries, physical fitness, body fat, growth & development, mental health, personality, etc. There are just too many effects to research in any one project.

Assume the researcher is taking a class in Exercise Physiology and has just become familiar with a hydrostatic weighing tank. The investigator knows that most pitchers aren't fat, but some are a little chubby. So the topic becomes "the effect of recreation baseball on male middle school pitchers as measured by a hydrostatic weighing tank." Yet further refinement is needed for greater specificity, because the particular kind of effect has not been identified, so the final project is titled " middle school recreation baseball and it's effect on body composition of male pitchers as measured by a hydrostatic weighing tank." If the title is too long for comfort, eliminate "...as measured by a hydrostatic weighing tank."

Other delimitations may include the following segmentation variables:

Demographic: age (Under 6, 6-11, 12-19, 20-34, 35-49, 50-64, 65 & up), occupation, education (grade school, middle school, some high school, high school grad, some college,

college graduate, post-graduate), sex, family size (1-2, 3-4, 5 & up), family life cycle (combinations of young, older, single, married, with children under 18, without children under 18, etc.), income (under $10,000; $10-$15,000, $15-$20,000, etc. to $50,000, $50,000 & up), race, religion, nationality.

Personal: height, weight, physique (somatotype), intellect (I.Q., SAT, ACT, etc. score), personality (compulsive, gregarious, authoritarian, dogmatic, etc.).

Social: social class (combinations of upper, middle, lower), political affiliation (democrat, republican, other), social distance.

Levels: beginner, intermediate, advanced, school grades.

Geographic: Region (East North Central; East South Central, West North Central, West South Central, South Atlantic, Middle Atlantic, New England, Mountain, Pacific), City size (under 5,000, 5-20,000, 20-50,000, 50-100,000, 100-250,000, 250-500,000, 500,000-1,000,000, 1-4,000,000, over 4,000,000), population density (rural, suburban, urban).

Use: Occasion (regular, special), user status (nonuser, first-timer, regular, potential user, ex-user), user rate (light, medium, heavy), loyalty (none, some, medium, strong, absolute), attitude towards service (hostile, negative, indifferent, positive, enthusiastic).

Finally, there are some practical considerations to make about the topic. Is the study feasible? That is, is the data readily available? Will the study be too time-consuming or too expensive? Will the study be of any pragmatic use? Will it satisfy formal requirements? What about timeliness? Are the necessary tools (computer, software, instrumentation, etc.) readily available? Are library resources adequate? Is transportation available if needed? Do I have a genuine interest and at least some background in the subject matter, or am I just trying to meet a requirement? A negative answer to any of these questions may mean a different study should be considered.

3. Developing Hypotheses

 hypothesis is an expected result. It may forecast direction, if there is adequate reason to predict it. For example, a researcher might expect a jogging program to decrease depression. HO (hypothesis): jogging 2 miles per day, 6 days a week will reduce mild depression. Or HO: daily aerobic conditioning will decrease mild depression.

There are times when the hypothesis direction may not be predictable, because it could go either way. For example: HO: there will be no difference in perceived comfort when reducing leg room in a motorcoach by one inch. In this case, one inch less leg room should not make much difference to the average passenger. However, leg room is already reduced to the lowest possible common denominator, and one inch could be a major factor in the comfort level perceived by a tour group. By stating the hypothesis in the *null* (no difference) form, the direction is not predicted and the hypothesis can be accepted or rejected based on data analysis.

It is understood the hypothesis must be testable. That is, there must be some way to obtain data and accept or reject the hypothesis. If there is no way to actually move the seats in a motorcoach so as to reduce the leg space by one inch, there is no way to test the hypothesis. In fitness circles, there is a theory that aerobic conditioning will increase the number of blood vessels in heart muscle. But how could the hypothesis HO: jogging 2 miles a day 6 days a week will increase the number of coronary blood vessels be tested, except by expensive, state-of-the-art instrumentation?

In some cases, it may be appropriate to use guiding questions instead of hypotheses. For example, an historical study may not be trying to support or refute hypotheses. That is, a researcher interested in the sport of basketball might use a question like "was Dr. James Naismith influenced by an ancient Mayan game using a rubber ball and iron rings when he invented basketball?" Such a question provides focus, yet the issue may be difficult to support or refute with quantifiable data.

4. Literature Search

A fter defining a problem, limiting it and stating hypotheses or guiding questions, the next step is to search the literature for related information in similar studies. A literature search can be manual, electronic or both. A manual search requires finding the information by hand, while the electronic search is able to use a computer to retrieve sources. Sometimes a combination of the two, such as InfoTrac is used. InfoTrac keywords retrieve printouts of sources; a manual search using filmstrip or a card reader is then initiated.

Regardless of the search method, a list of keywords, descriptors or identifiers are compiled, because many references list information by subject. Several keywords are necessary to start the process; others may be added as the search continues. For example, for a study about client characteristics other keywords that might be used are attitude, interest, visitor, participant, demographics, personality, traveler, etc. As the search continues, other descriptors listed in close proximity to the words being searched may be added for a more comprehensive list.

The purpose of the search is to find articles related to the research effort. A computer search will provide a print-out in alphabetical order. A manual search provides information about the source. Be sure to write down the *complete* reference as it is given, including, if applicable, author(s), date, edition (if book), title, journal or magazine, page numbers, city (if book), publisher (if book), volume (if periodical), issue number (if periodical) and any other information given. Frequently, the novice will omit some part of the bibliography and have to return at a later time to find the missing information.

Once enough references are recorded, find the source and read the abstract to see if the information appears relevant. Scan the article headings and selected paragraphs for pertinence, then read for content or discard if the information is not appropriate. Be sure to copy down keywords about the study on a note card that may be used in the research effort, along with any quotes, page numbers or abstracts necessary. For a few cents, most articles can be copied and kept for future reference. Be sure to copy all bibliographical information on the note card or article copy for future reference. Finally, look at the references given. Often, the bibliography will contain references overlooked until this time. When all the works cited begin to coincide with the reference list obtained, a thorough search has been done. A cursory list of books, indexes, other miscellaneous sources and periodicals where information about leisure, recreation and tourism can be found follows.

Books.   Books can be found in the following sources:

o   Books in Print. This source is a book itself containing a list of all books currently in print. It also has a list of publishers and their addresses.

o   Cumulative Book Index. This index has been published since 1898 (U.S. Catalogue) and includes all current books published in the English language. It is indexed by author, subject and title, and includes publisher, price and Library of Congress order number.

o   Book Review Digest. This digest, begun in 1905, includes reviews of books and lists price, publisher, and critiques.

o   "Travel Research Bookshelf". Several pages of new books, papers, and research listings, mainly in travel, tourism, hospitality and restaurants appear in the Journal of Travel Research published quarterly by the Travel and Tourism Research Association, Business Research Division, University of Colorado at Boulder, Campus Box 420, boulder, CO, 80309-0420.

Indexes: Some general indexes are:

o   Abstracts for Social Workers. Therapeutic Recreators may find some sections helpful, such as aging, alcoholism and drug addiction, and health and medical care.

o Completed Research in Health, Physical Education, Recreation and Dance. There are sections with subject headings, including a bibliography of published research, with a list of research abstracts.

o Dissertation Abstracts International. Doctoral dissertations are listed by author, subject and institution. These dissertations are microfilmed and available for a small fee.

o Education Index. This publication lists over 180 periodicals by subject and author. It contains all college textbooks published in the U.S, as well as publications of the National Education Association.

o Education Resources Information Center (ERIC). Computer or manual searches of two indexes are possible. Current Index to Journals in Education (CJIE), and Resources in Education (RIE) are listed by subject and contain EJ or ED accession numbers that are needed to find abstracted articles. Keywords called descriptors and identifiers are also published.

o Index Medicus (DIALOG MEDLINE). Listing over 2,000 periodicals from all over the world, it has a subject and author index in medicine and related sciences.

o Leisure, Recreation & Tourism Abstracts. A quarterly journal containing author, subject and geographical indexes. A computer search is available on DIALOG or ESA/IRS.

o Masters Abstracts. Contains quarterly abstracts of master's theses.

o Nutrition Abstracts and Reviews. These abstracts cover both animal and human nutrition.

o Psychological Abstracts. Contains topics such as drugs, human behavior, and human sexuality.

o   PsycINFO. This index is a computerized search of behavioral science literature by subject.

o   Reader's Guide to Periodical Literature. Reader's Guide was started in 1900 and since 1929 includes references outside the field of professional education. It contains both subject and author entries.

o   R.E.C.R.E.A.T.I.O.N. A spectrum of research papers which address societal challenges across the domains of leisure behavior. Includes completed research and methodical challenges to behavioral research.

o   Review of Educational Research. Summarizes research in the field of education and related areas.

o   Sociological Abstracts. Topics of interest may be demographics, human biology, and the sociology of health and medicine.

Other: Miscellaneous sources:

o   Business Statistics 1963-91. Superintendent of Documents, U.S. Government Printing Office, Washington, D.C. 20402. Includes tourism statistics.

o   Goeldner, C.R., & Dickie, Karen. (1980). Bibliography of tourism and travel research studies, reports and articles. Boulder, Colorado: Colorado Business Research Division, College of Business, University of Colorado. 762 pages. Contains information concerning recreation, travel and tourism.

o   Managed Recreation Research Report. A comprehensive resource of management information, ideas and trends for the recreation, fitness and leisure industry. Available from Lakewood Publications, 50 South Main Street, Minneapolis, MN 55402.

o   Moody's Industrial Manual. (Year) Contains some
     travel/tourism companies with balance sheets,
     descriptive narratives, etc. Enter with company name;
     ie, Disney, (Walt) Company, (The).

o   Moody's Transportation Manual. (Year). Contains
     financial and other statistical information about the
     transportation industry, along with specific financial
     information for individual business firms.

o   NRPA. (1970). Bibliography of theses and
     dissertations in recreation, parks, camping and
     outdoor education. Arlington: National Recreation and
     Park Association. 555 pages. An annotated
     bibliography of about 4,000 theses and dissertations.

o   NRPA. (1979). Bibliography of theses and
     dissertations in recreation and parks. Arlington:
     National Recreation and Park Association. This
     annotated bibliography updates the earlier version.

o   National Tour Association (NTA). P.O. Box 3071,
     Lexington, KY 40596-3071. NTA publishes Tours, and
     has a resource catalog that includes some free
     research studies and other research for a small
     charge. NTA is aware of current practices in the field
     and has a focus on practitioners (suppliers, operators
     and DMO's), and educators. NTA can plan and asist
     in researching most types of travel/tourism data.

o   Pisarski, Alan. (1985). An inventory of federal travel
     and tourism related information sources. Boulder,
     Colorado: Business Research Division, Boulder,
     Colorado. 107 pages. This inventory includes federal
     data programs.

o   Statistical Abstract of the United States (Year). Printed
     by the Superintendent of Documents, contains
     statistics for demographics, health, recreation and
     leisure, and travel/tourism.

o The Presidents's Commission on Americans Outdoors. (1986). A literature review. Washington: U.S. Government Printing Office. 165-816:64524. A review of research in outdoor recreation.

o Travel Reference Center, Business Research Division, University of Colorado, Boulder, Colorado. A clearinghouse that contains the largest collection of travel, tourism and recreation research in any one place in the United States. Computerized searches are possible.

o U.S. Industrial Outlook, (Year). Published by the U.S. Department of Commerce. Travel statistics are a part of the data given.

o U.S. Travel Data Center, Two Lafayette Center, 1133 21st Street, N.W., Washington, D.C. 20036. The center gathers, analyzes, publishes and disseminates travel research data. Each year the Center publishes the Outlook for Travel and Tourism, which is an analysis and prediction for the coming year.

o Walker's Manual of Western Corporations. Mainly financial statistics (balance sheets) on some travel/tourism business firms. Enter with firm name; ie, America West Airlines Inc.

Periodicals. Some selected periodicals to explore are:

Recreation and Leisure

o Adapted Physical Activity Quarterly.
o American Corrective Therapy Journal.
o American Educational Research Journal.
o American Journal of Health Promotion.
o American Journal of Sports Medicine.
o British Journal of Sports Medicine.
o Camping.
o Educational and Psychological Measurement.

o Employee Services Management.
o Journal of Health, Physical Education, Recreation and Dance.
o Journal of Athletic Training.
o Journal of Comparative and Physiological Psychology.
o Journal of Educational Psychology.
o Journal of Learning Disabilities.
o Journal of Leisure Research.
o Journal of Park and Recreation Administration.
o Leisure Sciences.
o Leisure Studies.
o Leisure Today.
o Parks and Recreation.
o Psychological Reviews.
o Psychology in the Schools.
o Quest.
o Recreation Management.
o Schole: A Journal of Leisure Studies and Recreation Education.
o Society and Leisure.
o Sociological Abstracts.
o Sociology of Sport Journal.
o Strategies.
o The Physical Educator.
o The Sport Psychologist.
o The Research Quarterly for Exercise and Sport.
o Therapeutic Recreation Journal.
o World Leisure & Recreation.

Travel and Tourism

o Annual Conference Proceedings (Yearly, TTRA).
o Annals of Tourism Research.
o Canadian Travel News.
o Courier.
o Festival Management and Event Tourism.
o Hotel and Restaurant Management.
o International Journal of Hospitality Management.
o International Tourism Quarterly.
o Journal of Hospitality and Leisure Marketing.

o   Journal of Travel and Tourism Marketing.
o   Journal of Travel Research.
o   Tourism Management.
o   Tours!
o   Travel Printout.

5. Collecting Data

After a literature search has been done, the next step is to collect data. The instruments used should be valid, reliable and objective. Prior to data collection, procedures for collection should be carefully planned. The type of statistical analysis should be determined, because the kind of analysis may depend on how the data is obtained and how subjects are assigned to treatments. If the study is experimental, provisions for controls employed are determined, with consideration made for internal and external validity.

Internal validity refers to how well the results can be attributed to the treatments used. If controls are not made for variables that could affect the data, the study may lack internal validity. Randomization of experimental subjects may assist in controlling effects from unwanted variables.

External validity is concerned with the generability of the results. Sample data should be representative of the population data if generalization is to occur.

6. Analyzing Data

Statistical tools are used to analyze quantifiable data. Prior to data collection, the type of analysis is determined. In a general way, the most powerful technique that meets data assumptions are utilized.

Many novice researchers are afraid of statistical tools. It is tempting to use techniques that are complicated and difficult to compute. Some novice researchers rely on computer print-outs without realizing what the data really mean. Advanced procedures are not the usual tools used in the literature. Reid and Andereck (1989) surveyed major tourism research literature and found the majority of analyses were descriptive. Specific procedures are found in later chapters of the text.

7. Reporting Results and Conclusions

ne way to report *results* is to list the data analysis as it occurred. That is, report what the data analysis revealed. Often tables, figures and graphs can be used for this purpose. Raw data can be placed in the appendix; only a synopsis is needed in the results section. Interpretation of the data may not be needed in the results segment, because conclusions can be drawn in the part that follows.

A way to reach conclusions is to return to the hypotheses or guiding questions and show whether or not they can be accepted or rejected by the data analysis. Further conclusions may be added beyond the analysis by attempting to explain why the results occurred as they did. Both quantitative and qualitative explanations are possible.

No attempt has been made to offer suggestions about organizing and writing the report. These topics will be covered in a later chapter.

Summary

How to research a topic has been presented in this chapter. The explanations are not intended as exhaustive; rather, suggestions are made for concept development of an overview of types of research and ways to research a topic. Some information sources given may be helpful as starting points in reviewing literature. Chapter 3 will discuss writing the report.

EXERCISES

1. Using **all** the sources listed in this chapter to find books, indexes, other (except NTA, Travel Reference Center & U.S. Travel Data Center) and periodicals, visit the library and determine if these sources are available. Turn in a typewritten list indicating where the sources can be found. For example:

	In Library			
Source	No	Yes	Location	Type[1]
Books in Print		X	1st Floor, East Room	Bound

[1]Bound, microfilm, microfische, call number (periodicals), etc.

2. Use *one* of the following keywords to obtain a printout from InfoTrac or a similar computer search. Print all references up to the first ten.

    Virgin Airways fare cut   corporate fitness
    recreation management   handicapped access
    hearing disabilities   forest trails
    Hard Rock Cafe   second hand smoke

3. Turn in a typewritten report on the subject of your choice using two master's theses and two articles found in either the Journal of Learning Disabilities, Journal of Leisure Research, Journal of Travel Research or Research Quarterly for Exercise and Sport. Each report should contain (if appropriate):

    Author(s)
    Date
    Title
    Name of journal
    Issue Number
    Volume Number
    Page numbers
    Type of research
    References (how many? Were all cited in text?)

Hypotheses (how many? Null, positive, negative?)
Methods or procedures used
Results
Subjects (how many? How selected?)
Statistical analysis (type? Confidence levels stated - .05, .01?)
Reaction (a brief statement of what you thought about the study).

4.   Select a topic you would be interested in researching. Turn in your topic title and explain in a typewritten paper how the topic was located and delimited. Include 25 references that could be used to research the topic.

Bibliography

Bookwalter, Karl W. (1965). Selection and definition of a problem. in M. Gladys Scott, (Ed.), Research methods in health, physical education, recreation (pp. 39-63). Washington: American Association for Health, Physical Education, and Recreation.

Carter, Edward. (1967, Spring Quarter). [Class notes taken in the course History and Philosophy, East Carolina University].

Goeldner, C.R., & Dickie, Karen. ((1980). Bibliography of tourism and travel research studies, reports and articles. Boulder, CO: Colorado Business Research Division, College of Business, University of Colorado.

Reid, Laurel J, & Andeereck, Kathleen L. (Fall, 1989). Statistical analyses used in tourism research. Journal of Travel Research. 21-24.

Chapter 3

Writing the Report

There are many styles of writing. Usually, colleges and universities adopt or modify a particular style to suit their needs. Campbell's Form and Style in Thesis Writing is a popular format. In a general way, leisure services publications often accept the American Psychological Association (APA) style, while tourism journals, such as the TTRA Journal, use the Chicago Manual of style. Although there are many differences in format, there are also many similarities.

The approach used in this text is intended to be general in nature. For style requirements, the student will need to check with the college or university department, while the prospective author can check the specific journal. If a journal issue does not contain submission requirements, write to the journal editor for style conditions.

Whether writing a thesis, dissertation or research article, the basic form is similar. The discussion that follows will focus mainly on general thesis procedures, but many of the parts, while not described as chapters, apply to journal writing as well. The descriptions will center around preliminary pages, chapter organization and appendices. Specific points about style, such as levels of headings, pagination, reference format, etc. will not be discussed because they vary with the department or journal requirement.

Preliminary Pages

The preliminary pages usually include elements such as a title page, approval page, acknowledgement, abstract, table of contents, list of table, list of figures and copyright.

Title Page. Although there are many variations of title page format, including numbers of spaces between each element, a title page usually has the title of the research, the authors full name, highest degree received, college or university and full date. The title should be brief, perhaps no longer than 15 words, but should accurately portray what the topic is all about. Abbreviations are not usually listed in a title. A title page is appropriate for all formal writing.

Approval Page. An approval page is required for a thesis or dissertation. It may contain the title, author and various signatures, such as thesis advisor, committee members, graduate dean, etc. Spacing and position of the page elements vary with style.

Acknowledgement. The acknowledgement is reserved by the author for anyone to whom the author feels indebted to in motivating or developing the research. Sometimes it may be the major advisor, if the person has motivated the author *beyond what is normally expected.* At other times it may be a person who was not acting in a formal capacity but nonetheless provided exceptional help. Acknowledgements may also be made for permission to use and publish instruments, tables or other copyrighted or patented materials. Often a family member has provided the motivation. However, an acknowledgement is not necessary, and should be reserved only for those persons who have provided *exceptional* assistance, motivation or special use of instruments or materials.

Abstract. The abstract should be brief. It may contain up to 150 words or more as requested by the editor. The abstract should contain the title, a very brief description of method or procedures, instruments or data-gathering devices, subjects, results, statistical levels examined and conclusions. It is important to spend a lot of time organizing words in the abstract, because even though it is very brief, many people will read it and decide to continue or stop based on the content.

Table of Contents. Although not used for a journal article, the thesis or dissertation, and even term papers often require them. The heading Table of Contents is usually appropriate, followed chronologically by the major headings and page numbers. The use of a period symbol (. . .) every other space from the subject listed to the page number is a technique that may improve readability.

List of Tables. The heading List of Tables followed by a chronological listing with page numbers at the bottom of the page in arabic is included in preliminary pages. Often, a table is numbered by the chronological digit as it occurs in the chapter. For example, the first table in chapter 1 may be numbered 1.1, while the second table in chapter 2 would be 2.2, etc. Tables may also have numerals that progress consecutively, starting with the first table published through the last table shown, without reference to the chapters; ie, table 1, table 2, etc. A table is normally titled and numbered just above the *top* of the table. Different styles follow spacings and positions of the table heading according to the particular style.

List of Figures. A list of figures follows the same procedures as the list of tables. Usually, figures are numbered continuously as with tables and titled at the *bottom* of the figure.

Copyright. An *automatic* copyright exists when the document is published. However, to protect the copyright, it is best to formally register the document with the U.S. Copyright Office, Library of Congress, Washington, D.C. 20559. Forms are available from the copyright office. Form TX should be requested for published and unpublished nondramatic literary works.

The Copyright Act has a mandatory deposit requirement for works published in the United States. For registration, two copies of the document must be sent within 3 months of publication. A copyright fee of $20.00 and a filing fee (slightly less) was needed in 1994. A librarian can assist in getting the copyright secured for theses, dissertations and other works. Different forms and fees may apply for different works.

Chapter Organization

This section will focus on how a research paper is usually organized. Although a journal article will not have chapter headings, the general headings follow a similar format. It should be understood that most types of research can follow the headings presented here. However, different types of research may not have all the headings listed. Also, different departments, journals and types of research may require a different format. This discussion is not intended to be all inclusive, but serves as an introduction of ways to organize many forms of writing.

Basic procedures. Write in past tense. The research, when

completed, describes what has happened. The literature review portrays information that has already been written.

Double space all text. A committee chair, committee member, journal reviewer, editor or the author may need space for comments helpful during the revision process. Quotations are a special case and should follow specific procedures mentioned in the style manual.

Have an organizational structure. Unify the topics discussed by helping one topic flow into another one without an abrupt change of thought, when possible.

Keep it simple without "talking down" to the reader. Use of too many technical, scientific words may discourage a person from reading the paper and can hide the meaning of what happened. How many technical papers have you read and wondered when you finished what the author really said? Such a paper may be technically correct, but a major purpose is to communicate what happened to someone else so that *they* can *understand* it.

Use the recommended style methods for referring to work cited in the body of the text. Most styles do not require footnotes for every work and author. A simple way found in both leisure, recreation and tourism literature is to say Jones (1994), or "this method (Jones, 1994)," or "... found this method to be highly effective (Jones, 1994). The complete reference can then be found in an alphabetical listing at the end of the chapter.

The length of a dissertation, thesis or journal article vary with the content and place it is published. The author has seen a satisfactory thesis of around 35 total pages, although many vary from about 50 to over 100 pages. The length of a journal article varies with the journal, and is usually prescribed by the editor along with the style requirement.

After selecting and delimiting the topic, the next step is to begin work on developing the study. Usually, the subject headings in the first chapter help to clarify the project. By completing chapter one, the researcher is ready to investigate the literature, which may help to further refine the study.

Chapter 1. INTRODUCTION

T he first chapter or heading is usually titled Introduction. The discussion that follows concerns section headings that are generically appropriate for most first chapters. The use of quotes or a lot of references is not as appropriate here as in other chapters. After all, the research is what the author has done, not what others did.

Introduction

Write a short introductory paragraph. The purpose is to get the reader's attention and introduce the topic to the reader. An introductory paragraph may deviate more from formal writing than any other paragraph in the paper. Write towards the audience the research is designed for. For example, one introductory statement was "Coaches and swimmers alike are often baffled by swimming performance when top-flight competition is held." (Shockley, 1971). The audience was swim coaches and swimmers, and that opening statement gained the attention of that particular audience because it was of interest at that time.

Rationale

The rationale is the reason for the study. Why is the study needed? Why is the research important? What is the general background for the study? Answering these questions in one or two paragraphs establishes a need to complete the study. Editors, reviewers and committee members will need to be convinced the rationale is appropriate before the study can proceed further.

Problem Statement

A clear statement is needed here. "The purpose of this study was to..." may be an appropriate start. The statement should be simple and concise, yet portray in one or two sentences exactly what the specific subject or immediate action was.

Hypotheses

A hypothesis is a logical explanation of what is expected. It is not based on whim or fancy, but is a theory formulated from past experience, known results or prior research. Sometimes, guiding questions are used in place of the hypothesis.

The hypothesis usually states a relationship between variables. It must be testable, as mentioned in Chapter 2. An example of a null hypothesis (no difference) was given in Chapter 2. Examples of hypotheses worded in a positive direction might be:

Clients will show a preference for tour itinerary B over itineraries A, C and D.

Patients with mild depression will show improvement in social distance over patients with severe depression.

Trail A will show more erosion than trails B and C.

Hypotheses can also be written in a negative way, such as:

Male swimmers will show less improvement in time for 100 yards free during the six week training period than female swimmers.

There will be fewer telephone inquiries requesting information when using marketing plan 1 than when using marketing plan 2.

In each case, the hypotheses are composed in a simple way and they can be tested, either with known instruments and/or simple statistical techniques. A more detailed discussion of hypotheses is found in Chapter 2.

Definitions

Definitions of ordinary words in which there is no dispute over meaning by professionals in the field do not need to be redefined.

Where there can be different interpretations of a word meaning, the author should make clear the definition that was used. For example, most fitness professionals would know what is meant by the word *body fat* or *body composition*. By the same token, most practitioners would know what *skinfolds site* measurements are. However, when using skinfolds measurements, there are various places and combinations where these anthropometrics are taken on the body. Whereas it is not necessary to define body fat, it is important to define skinfolds site. Or a visit could mean any number of things, so the researcher should define what is meant by a visit. It is particularly important to define all ambiguous major terms mentioned in the title and problem statement.

Scope of the Study

This section may be several paragraphs long and is a general overview of the study, including a gross discussion of subjects, procedures and instruments used. Whereas it is not as detailed as Chapter 3, it should leave little doubt as to how the study will be done.

Limitations

Limitations are sometimes described as (a) limitations and (b) delimitations. In this context, limitations are shortcomings or influences which cannot be controlled by the investigator. For example, when using human subjects, it is virtually impossible to control every aspect of their behavior - diet, amount of sleep, etc. When testing performance, how can it be ascertained the subject gave a maximum effort, or answered a questionnaire truthfully?

Delimitations refer to such things as choices made by the investigator - the use of one instrument over another (is one instrument really better than another?), restricting the number of questions in a survey because of time constraints, eliminating a data site due to the cost involved, etc. These kinds of choices do affect the study, and recognizing and stating how the choices made affect data gathering and analysis is important.

Sometimes the two categories mentioned are combined into the single heading of limitations. There are many variations in the literature, and the researcher will need to follow departmental and major advisor advice. Current practice is to identify *major* shortcomings so that the investigator and readers will be aware of them.

Significance

The significance refers to the probable value of the study. Identify not only to whom the study is important, but also in what way and how. This paragraph is different from the rationale, which has explained why but may not distinguish who can use the information, in what way and how.

After the last topic is discussed, write a very brief paragraph summary, and describe in one sentence or less what topic the next chapter will address.

These generic headings, their derivatives or combinations are usually found in most introductory chapters. Some headings may not be necessary in a particular study, while additional headings, such as specific objectives, may be appropriate. In the case of college or university students, the department and/or major advisor will prescribe what is needed. For a journal article, read what other authors have done for ideas.

Chapter 2. REVIEW of LITERATURE

A review of literature can be a time-consuming and tedious process. The objective is to write a summary of the research already accomplished and/or related to the stated problem. Usually, a bibliography is obtained and the articles then examined for pertinence. Copies of the articles can be obtained or 5" X 8" note cards may be used. All bibliographical data should be copied on the note card or article copy for future reference. There is an unwritten axiom that failure to record any part of the reference means the author will have to return to the source in order to use it later.

Major information that needs to be listed includes the purpose of the study, description of subjects (if used), methods and techniques used and findings and conclusions. It is also wise to note any quotes and page numbers for them that might be employed later. Also, page numbers of particular points of interest may prove helpful later. The attempt is to write down all information that might be needed at a later time to avoid having to return to the source for additional information.

As the review proceeds, note the references given for every article. These sources are origins for additional information about the topic. When the bibliography begins to repeat what the author already has reviewed, most of the references have been found. Chapter 2 contains additional discussion regarding a literature search that would be worth reviewing at this time.

Students sometimes ask, how many years back do I need to go? There is no definite answer to this question. My major professor told me to get information for a dissertation as far back as could be explored! But students or researchers with limited time can't possibly review *all* the literature, even with computer technology. Anyway, of what use is information several decades old? The answer to this question depends on the topic researched. Current information may be more valuable, but findings and ideas recorded a century ago may be applicable today, again depending on the topic. Start with the last 20 years. As the review proceeds, it will become readily apparent how time-consuming the research will be. It may also surprise the investigator how necessary it is to go back even further. While completing a dissertation, one small part of the study explored athletics and academics. The author was surprised to learn that in general educational literature, athletes had higher grade point averages and national test scores when compared to general students even prior to 1920. In fact, Aristotle, Socrates and Plato all advocated high fitness levels for optimum human development. One answer as to how far back to go is as far back as time and resources will allow.

The literature review will reveal information that can be organized into topics. These topics become headings or points of discussion. Often, several researchers have reported essentially the same findings. In this case, combine the findings and reference all research related to the *same subject* in the same sentence. For example, some prominent coaches who used isometric exercises (*same subject*) were: Clotworthy (1964), Counsilman (1968), Daland (1964), Gambril (1969), and Stager (1966).

All studies closely related to each topic should be cited and a brief summary of each study presented. Compare and contrast each study to the present study, if appropriate.

It should be apparent that not all references reviewed will be used in the present study. Eliminate the research that does not apply. How many final references should a paper contain? Certainly dissertations will usually have more works cited than a thesis, and a thesis may have more citations than a journal article. Depending on the topic, perhaps a number to start with is 50 or more references for a thesis. Again, there is no magic in any number of references. Cover the topic and convince the reader that a thorough review has been done, and the number will take care of itself.

After the literature review has been concluded, write a short summary of the findings and introduce the next chapter, which is methods and procedures used.

Chapter 3 METHODS and PROCEDURES

Introduction

Describe briefly what methods and procedures were used in the study. Include the type of research, such as analytical, experimental, historical, etc. Include the *general* nature of the scope of the study, but reserve a detailed description for the next section. The objective of this section is to acquaint the reader with the way the study was organized to collect data. Advisors, committee members, journal reviewers and editors look closely at this section to see how well the research has been developed.

Procedures

Define each step in the process that was followed, including subjects, instruments and procedures. Include, if appropriate, a detailed discussion of techniques, such as research design; kind of instruments; calibration of instruments used; how subjects were selected; description of subjects; randomization procedures; controls utilized; considerations such as development of measurement tools (validity, reliability, objectivity group bias, etc.); explicit discussion of methods followed, step by step, including a projected calendar; statement regarding risk to human subjects, with a copy of same in the Appendix; detailed discussion of analysis to be used, such as statistical, and other plans. A black and white glossy photograph showing instruments or data collection from a subject or subjects is invaluable. Submit a separate page for a photograph to a journal.

To conclude the chapter, summarize and introduce the next chapter, which is analysis of the data.

Chapter 4 ANALYSIS of DATA

Usually, the data analysis describes what was determined quantatively or qualitatively without discussing the reasons for what occurred. Just write what the data analysis revealed. Tables, figures and narrative discussions may be used. Try not to include a table and narrative or figure that describe the same data. Use a table instead of a long, drawn out narrative. Tables are employed to present large amounts of data for descriptive or comparative purposes. Large amounts of data can be presented using *grouped* data, if necessary. Rounding numbers in a table often makes the data more readable (original numbers can be placed in the Appendix). Most journals require tables and figures to be submitted in a special format on a separate page. Multiple copies may be required.

When reporting statistical data, be sure to include (APA, 1983):

> ...magnitude or value of the test, the degrees of freedom, the probability level, and the direction of effect. Be sure to include descriptive statistics (e.g. means or standard deviations)...

Use a summary paragraph at the end of this section and indicate what the next chapter, Results and Conclusions will decsribe.

Chapter 5 RESULTS and CONCLUSIONS

This chapter begins by summarizing the results, explains what the data means, and gives recommendations for further study. In a journal article, this section may be combined with data analysis.

Results

Explain what the data mean. Use the hypotheses or guiding questions as a framework for organization. When accepting or rejecting hypotheses, explain why the result happened. Whereas speculation should be avoided, reasons that have sound theoretical basis should be formulated.

Recommendations for Further Study

After the research has been completed, the author is in a unique position to make recommendations for further research. What new variables need to be introduced? Would a different treatment be advisable? Could a different level be explored? Should the present study be refined or changed in any way to explore different result possibilities? Is there anything the author found during the research investigation that needs examination?

BIBLIOGRAPHY or REFERENCES

Generally, a bibliography refers to all works reviewed and additional sources that are important, even if not cited in the text. References are all of the works cited in the text. The format may be recommended by the department, major advisor or journal. If a bibliography is used, be sure to include all works cited in the text as a part of the bibliography. List each citation alphabetically. Spacing and ways of listing the information vary according to style. Works cited are usually single spaced, as opposed to double spacing in the body of the text. One way to organize this section is by books, periodicals and other.

Books

Depending on style, there are different ways to list these types of references. Some styles require capitalization of every word in the title, others do not. Some styles underline spaces in the title, while others do not. There are different ways to list more than one author. Use the way recommended with the specific style.

Periodicals

Periodicals have different ways of listing information too. Citing journals, magazines, newspapers, etc. with one author, two authors, three authors, etc. each have different elements recorded in different places. Check the appropriate style manual or journal .

Other

There are other kinds of references, such as government publications, theses, interviews, electronic media, film, recordings, etc. that require different reporting format. As in all instances, use the method suitable for the style.

Often there is no categorization; bibliography or references are simply listed in alphabetical order.

Appendix

The appendix contains materials for which no place is found in the body of the text. Whereas journal articles rarely have appendices, dissertations and theses make extensive use of them. Examples of material suitable for the appendix includes but is not limited to a human subjects approval form, instructions to subjects, copies of original data, paper instruments used (test, questionnaire, sociogram, social distance scale, etc.), cover letters, sanction authorization, etc. Each different piece of material is a different appendix. Appendices are usually numbered by uppercase letters, such as Appendix A, B, C etc. A sheet of paper with the word Appendix centered and all works listed in that appendix usually precedes data in each appendix.

Use a Computer or Word Processor

In order to learn to write quickly and accurately, learn to use a computer or word processor. Technology today has made the days when constant revision on a typewriter unnecessary. A word processing software package combined with a spreadsheet gives all the equipment necessary for creating and saving written documents.

Word processing software gives the writer the capability of desktop publishing. Just write and save. Type in the margins and they appear on every page. Give instructions for pagation and it occurs on every page. Looking for a synonym? Access a thesaurus. Want to be sure of spelling? Use the spell checker. Create tables? That function may be there. Unsure of grammar? Even grammar can be checked. Want to insert or move information? Try cut and paste. Need to find a word or phrase in a long paper? Try search. Want to know what the document will look like before it is printed? Try preview. Create an equation? It can be done with special graphics functions. There is no way to list all the advantages of commercial software, but there is no excuse for not being able to create a professional looking paper with the technology available today.

Worried about drawing graphics? Try a spreadsheet. Often just typing two columns of numbers and accessing a graph function will show the graph *already* created. All that is necessary is to add headings, labels, and other custom features. Afraid of making a mistake with numbers? Learn to use a spreadsheet, type in cell procedures (formulas) and watch the computer give results in micro-seconds. Create a graph in one spreadsheet, move it to a word processor, and even have it updated automatically when different numbers are used. Every word, graph and table in this text was written, typed and checked by the author, who had no special training.

A decade ago, this technology was not possible. Many a student had to take their paper to a typist for a final copy. Ever get two chapters of a dissertation typed, then have the typist tell you she can no longer continue, and find every other typist in town too busy to accept new clients? All of the typing problems can be eliminated with just a cursory knowledge of computer basics. Save the copy and it can be revised or retrieved and printed at will.

<u>Epilogue</u>

Of all the difficulties in writing, a very important consideration is to write clearly and simply so that the document can be understood. Omitting words or using ambiguous words and phrases can lead to a misunderstanding or misrepresentation, regardless of intentions. The following quotations show how writing can be misunderstood (Van Buren, 1985).

> (Monday) "FOR SALE-R.D. Jones has one sewing machine for sale. Phone 948-0727 after 7 p.m. and ask for Mrs. Kelly who lives with him cheap.

> (Tuesday) "NOTICE-We regret having erred in R.D. Jones' ad yesterday. It should have read: One sewing machine. Cheap. Phone 948-0707 and ask for Mrs. Kelly who lives with him after 7 p.m."

> (Wednesday) "Notice-R.D. Jones has informed us that he has received several annoying telephone calls because of the error we made in his classified ad yesterday. His ad stands correct as follows: FOR SALE-R.D. Jones has one sewing machine for sale. Cheap. Phone 948-0707 p.m. and ask for Mrs. Kelly who loves with him."

> (Thursday) "NOTICE-I, R.D. Jones, have no sewing machine for sale. I *smashed* it. Don't call 948-0707, as the telephone has been out. I have not been carrying on with Mrs. Kelly. Until yesterday, she was my housekeeper, but she quit."

Summary of Chapter 3

Some generic ways of writing a report have been mentioned. General tips about writing, style, chapter organization and chapter headings were discussed. Although not exhaustive, the procedures mentioned are a starting point through which a study may be developed. Chapter 4 will discuss descriptive research.

Exercises

1.   Using the topic selected and delimited from Chapter 2 Exercises and the headings listed in this chapter, type a rough draft of Chapter 1. Use the style requested by your advisor or department.

2.   Using the topic and references completed in Chapter 2 Exercises, explore the references and type a rough draft of Chapter 2. Use the style requested by your advisor or department.

3.   Using the topic already developed from Chapter 2 Exercises, type a rough draft of Chapter 3. Use the style requested by your advisor or department.

4.   Find 50 references on a topic you would like to research. Type a list of these references in the style requested by your advisor or department.

References

American Psychological Association. (1985). Publication manual of the American Psychological Association (3rd ed.) (p. 27). Washington, DC: American Psychological Association.

Shockley, Joe M. Jr. (1971). An Analysis of Performance of the Swimmers in the 1971 NCAA University Division Championships, With a Description of Personal Variables and Training Methods. Unpublished doctoral dissertation, The University of Georgia, Athens.

Van Buren, Abigail. (1985, March 27). No title. The Caledonian-Record. p. 13.

Chapter 4

Introduction

Most research approaches are the same. Basic procedures were given in Chapter 1. These steps are the general methods to use for the particular type of study investigated. Further insight was provided by steps in the scientific method in chapter 2. Using these strategies to plan the inquiry is one way to organize the research process.

The major differences in types of research lie in collecting and analyzing the data. Since elements of the descriptive research process have already been discussed in Chapter 1, this chapter will focus on refining a problem; stating assumptions and limitations; and developing, collecting and analyzing the data. Writing the report should follow the guidelines given in Chapter 3.

Descriptive Research

Questionnaire Research

In questionnaire research, the investigator uses an instrument already designed or develops a questionnaire to gather data. This type of research is used frequently in tourism, leisure services and recreation.

There are many questionnaires already developed that can be used or modified with permission. For example, NTA has questionnaires in the field of travel/tourism. The Journal of Travel Research contains questionnaires. Dissertations and theses are sources for questionnaires in leisure and recreation. Some books, such as Jubenville (1976) *Outdoor Recreation Planning*, Corbin (1970) *Recreation Leadership*, business texts, sociology books, etc. contain instruments that may be appropriate. One difficulty with these tools is that published data about validity, reliability and objectivity are often missing. By assuming face validity, using published directions that imply objectivity and checking for reliability, most problems are overcome.

<u>Steps</u>. Steps in questionnaire research include defining the problem, determining assumptions/limitations, developing the questionnaire, gathering the data and analyzing the data.

<u>Defining the problem</u>. The first thing to do is determine what problem to study. Start with your own ideas, then review the literature for others. Chapter 2 reviewed how to discover and limit a topic to study. Possibilities for study include (see Chapter 2 for more):

o   Occupancy rate (load factor). Number (average) occupied/total number. Total number may refer to number of rooms; seats on a motorcoach, train, airplane, etc.; seats/tables in a restaurant; maximum number of people (room, gym, pool, tennis court, grandstand, etc.); and similar data. For percentage, multiply by 100. For example, the occupancy rate of a motel with 100 total usable rooms for a single night with 75 occupied is 75/100 X100 = 75%. The data may be more appropriate when compiled for a month, season, year, etc., using average ÷ maximum available.

o   Simple comparisons. Comparisons may be percentages or other statistical analyses, such as relationships that are discussed later. Some possible comparisons could be:

Table 4.1. Examples of Simple Comparisons.

Measure 1	vs. Measure 2
Number of clients registered	Peices of mail answered
Snowfall (inches)	Number of clients on ski lift

Number (or type) of event/attraction	Number of tourists
	Income (gross or net)
	    company store
	    hotel
	    restaurant
	    bar

Income (gross or net)	Inflation rate Oil price
Residents	Travellers

o   Miscellaneous client data.

Age breakdown.
Martial status, family make-up, etc.
Length of stay.

Satisfactions, dissatisfactions.
Mode of travel, movement of people.
Reasons for travel:
    environmental.
    economic.
    sociological.
Trip characteristics:
    socioeconomic.
    opinions, attitudes.
    origins, destinations, places visited.
    travel purposes.

Determine assumptions/limitations. *Assumptions* are facts taken for granted that are not studied or controlled by the researcher. These facts are thought to be common knowledge. Assumptions may or may not be true, but they alert the reader as to facts the researcher believe are true. Some examples of assumptions are:

o   The client desires exemplary service.

o   Price range is (is not) a major factor..

o   The client is in good general health.

o   The client is willing (unwilling) to spend more than 1 minute to answer questions.

o   The client can read.

Limitations are shortcominngs or influences which cannot be controlled by the researcher. Some examples are:

o   The mode of transportation to destination.

o   Personal characteristics of clients.

o   An accurate public database.

The reader is referred to Chapter 1 for further discussion about limitations and *delimitations*.

Develop the questionnaire. A questionnaire study must be practical. Some practical considerations are:

a.   time. How long will it take to gather, analyze and interpret the data? If this process takes too long, the topic may not be practical.

b.   cost. If there is not enough money to complete the project, or if the cost is prohibitive for any reason, the study may not be feasable.

c.   generalizibility. How closely will the data represent the population? Scientific sampling procedures will make the data more useful. The concepts of validity, reliability and objectivity applied to questionnaire design will assist in generalizing data to the total population for which it is designed.

      Validity. Validity refers to whether or not the questions are the ones that *should be* asked. Some ways to establish validity are:

      o   Review the literature. What items related to the topic are being studied by others? What are other people doing? What are other people saying?

      o   Examine related sources (see Chapter 1). For example:

            State Comprehensive Outdoor Recreation Plan (SCORP).
            National Outdoor Recreation Plan (NORP).
            Books
            Periodicals
            Infotrak & other electronic databases.

      Your own ideas and experiences starts the process. Look for ideas, items and questions that interest you. Start with **construct validity** - validity that by the very design of the questionnaire itself defines what questions should be asked. Then make a rough draft of the instrument based on your ideas and research. Be sure the draft is simple and objective, a point discussed briefly later.

Once the draft is completed, submit it to a panel of experts. The researcher defines who the experts are, but it is clear these people must be knowledgeable about the topic. The panel probably does not meet as a group in the same place at the same time, but their collective review establishes expert scrutiny. Examples of possible panel experts may include:

State SCORP writers.
City recreation department managers in a region or state.
Chambers of commerce.
State or national professional committee officers.
NTA Research Committee.
NESRA Research Committee.

Ask the panel if they recommend any changes. Try to eliminate bias in charging the panel. What if you asked, "would you **add** any questions?" Think of all the additions you might get! Focus the panel on changes, rather than add/deletions.

Finally, make the changes a majority of the panel recommend. If there is not a majority recommendation, accept the draft as written.

Reliability. Reliability refers to how consistent the answers obtained are. If different answers are given by the same person at different times, which one is correct? The usual way to check this trait is to ask a sample group in a pilot study to answer the questionnaire on at least two separate occasions closely related in time, perhaps 6-24 hours apart. If too long a time passes before obtaining second answers, conditions may change to such a degree that the answers are truthfully different.

One way to administer a test-retest method is to take the validated questionnaire and define the population for which it is intended. Then obtain a small sample of this total population (discussed in a later chapter). Ask the respondents to complete the questionnaire without mentioning the second request. Be sure to obtain the name, address and phone number of the person involved, but insure confidentiality of same. Then 6-24 hours later, ask each person to complete the form again. Be careful how this second administration is done. If the person is made aware the first and second questionnaires will be compared for consistency, the results may be adversely influenced. One way around this difficulty is to simply say, "scientific procedure requires that I ask you to complete this questionnaire again. Your help and understanding will be really appreciated."

Be sure to mark which questionnaires were first and which ones were later, so that all the first ones can be compared with the second group. Reliability may be found by using a correlation or ANOVA statistical technique, which are mentioned in later chapters. If the researcher believes the data is difficult to compare statistically, a simple percentage of agreement is possible. This technique uses the first administration of a question as the total, and obtains the percentage by dividing the number who answered the question the same the second time by the total number answering that question the first time. For example:

Table 4.2. Number of Respondents Answering the Same Questions Twice.

Question Number	Number Answering First Time	Number Answering the Same Later
1	31	31
2	30	25
3	30	20

For question 1, the percentage of agreement is found by dividing the number of respondents answering the question the same the second time (later) by the total number who answered the question the first time. That is:

$$\frac{31}{31} \ X \ 100 \ = 100\%.$$

For question 2, the data reveal a percentage of agreement of:

$$\frac{25}{30} \ X \ 100 \ = 83.3\%.$$

For question 3, the percentage of agreement is:

$$\frac{20}{30} \ X \ 100 \ = 66.7\%.$$

There is no known rule as to what establishes a satisfactory percentage of agreement. Obviously, when the percent is high, the question is reliable. When the percentage is low, the question is not as reliable. The author suggests revising or eliminating any question with a percent lower than 80. Thus figure may seem low, but experience over a lifetime has shown it may be appropriate. Most questions will be in the 90% or higher range.

Using this method to determine reliability, each question **and** the total questionnaire can be tabulated.

Objectivity. Objectivity refers to how well different people administering the same questionnaire obtain similar results. Sometimes a person can unintentionally bias answers by words, mannerisms, gestures, definitions of words, explanations to respondents, etc. To help eliminate this type of bias use the following techniques:

o   Keep words simple.

o   Keep directions simple and easy to follow.

o   Include word definitions when necessary. They could be printed on the back side of a one page questionnaire.

o   Use consistent procedures in questionnaire design. For example, keep the same type of questions (all yes-no questions; same kind of Likert Scale, ie, categories of 1-5; 0-7, etc.; same type of directions throughout). Use the same variety of answering procedure for all questions (circle all answers; check all answers, etc.).

Objectivity can be determined by having different people administer the questionnaire to a sample group of people and compare the data through correlation, ANOVA or percentage of agreement techniques.

Group bias. Questions can have different meanings to different groups of people. This phenomenon occurs when different sub-cultures give different meanings to words or emphasize words differently. The researcher can check to see that different social, religious, ethnic etc. groups can answer the questions consistently by using a pilot study of the desired group. A comparison of groups under study will then determine if group bias is present.

A different type of group bias may occur as a result of sampling techniques. This type of bias is enhanced when groups are chosen as a basis for random selection and individuals from the groups are selected to represent the group. Latham (1991) describes this type of group bias and suggests increasing the number of clients interviewed (N) as a way to bring this error within acceptable limits.

Gather the data. Once the questionnaire is scientifically designed, the next step is to gather the data. Use sampling procedures as described in a later chapter. To enhance the number of returns, try some of the following techniques:

o   Have the questionnaire printed on quality paper by a commercial printer.

o   Get a sanction for the study. A sanction is a **short** statement by a professional group or person that the study is worthwhile and needed. For example, the author once completed a study of NCAA swimming. The Education and Research Committee of the College Swim Coaches Association of America was asked to sanction the study (the same group was used as a panel of experts for validity inference). A cover letter, signed by the chair of this committee on letterhead stationery was included in the mailing. The researcher will need to determine who or what organization could add impetus to the study through a sanction.

o   Enclose a self-addressed, stamped envelope.

o   Agree to share a brief summary of the results with the respondent. This communication can be in the form of a separate mailing or published in a professional journal or newsletter the respondent receives.

o   Limit the questionnaire design to one page, two pages maximum. People just don't have time or interest to answer a long series of questions.

o   Include a time deadline, perhaps within two weeks. A deadline that is too short or too long will lower returns.

o   Personally collect the data at a conference or workshop instead of mailing.

o   Follow-up. Follow-up measures are important. Be sure to keep track of the data as it returns. After the deadline, try some of the following techniques to get any missing data:

Phone and complete the questionnaire right then.
Send a second mailing. **Enclose a dollar bill.**
Fax a second request.
Ask sanctioning personnel to write or phone.
Ask a friend of the respondent to contact her/him.

Analyze the data and report the findings. Data analysis is covered in later chapters, whereas Chapter 3 discussed writing the report.

Scaling. Chapter 6 discusses the general concepts of scales; ie, nominal, ordinal, interval and ratio. Scales that are nominal and ordinal require the use of non-parametric statistics for analysis and interpretation, whereas interval and ratio scales may use both parametric and non-parametric techniques. The literature is in disagreement as to whether or not using words to represent a number allow measurement beyond the ordinal scale. One authoritative source (Seigel, 1956) believed scales used by behavioral scientists in areas such as attitude measurement are ordinal at best.

Many scales are designed to measure client attitudes. Major types include comparative, Likert, semantic differential, and Thurstone. Although examples shown will *usually* have scale values starting from low to high, left to right, some researchers claim it is best to alternate the values from left to right in order to reduce what is known as the *halo effect* where a client may tend to check one side or the other.

Comparative scales. Comparative scales may be used for determining client preferences and attitudes. They are also used by tour planners prior to selecting a particular carrier or supplier.

Table 4.3. Examples of comparative dara.

A. Leisure/recreation	Poor	Neutral	Good
1. Program 1 vs. program 2.			
2. Site 1 vs. site 2.			
3. Leader 1 vs. leader 2.			
B. Sport Management	Boring	Neutral	Not Boring
1. Marketing technique 1 vs. 2			
2. Schedule 1 vs. schedule 2			
3. Equipment 1 vs. 2			
C. Travel/tourism	Not Noizy	Neutral	Noizy
1. Hotel 1 vs. hotel 2			
2. Mode 1 vs. mode 2			
3. Campsite 1 vs. campsite 2			

Likert scales. There are many forms of Likert scales. One form uses a five-point rating where clients check a five-point scale as to whether they disagree strongly, disagree, are undecided, agree or strongly agree (Likert, 1970). A variation is to use wording relating to importance, such as not at all important, not very important, somewhat important, very important and extremely important. Ratings of very poor, poor, fair, good and excellent may also be used.

Table 4.4. Examples of Likert scales.

A. Leisure/recreation	Strongly Disagree	Disagree	Undecided	Agree	Strongly Agree
1. I like to jog in the afternoon.					
2. I like to relax in the park.					
3. I am afraid to walk alone in the park.					
B. Sport Management					
1. The facility is always clean.					
2. Newspaper accounts of my team are adequate.					
3. My team is scheduled fairly.					
C. Travel/tourism					
1. My hotel room is fine.					
2. I like a lot of free time on tour.					
3. My tour guide is always helpful.					

Semantic differential scales. These scales are designed to measure attitudes with words that mean the opposite, usually on a 5 or 7 point scale. Word pairs are first determined and the scale constructed. Examples of word pairs follow (Osgood, C.E.; Suci, G. & Tannenbaum, P., 1957).

o	active/passive	o	angular/rounded
o	beautiful/ugly	o	calm/excitable
o	colorless/colorful	o	cruel/kind
o	curved/straight	o	false/true
o	good/bad	o	hard/soft
o	important/unimportant	o	masculine/feminine
o	new/old	o	savory/tasteless
o	slow/fast	o	uncessful/successful
o	untimely/timely	o	usual/unusual
o	weak/strong	o	wise/foolish

Using the same idea, word pairs are determined by the researcher. An example of one form of semantic differential scale follows.

Table 4.5. Examples of semantic differentials.

A. Leisure/recreation						
Joe Bannon park is usually:						
1. littered	_____	_____	_____	_____	_____	not littered
2. safe	_____	_____	_____	_____	_____	not safe
3. fun	_____	_____	_____	_____	_____	boring
B. Sport management						
The sports complex is usually:						
1. cold	_____	_____	_____	_____	_____	hot
2. filthy	_____	_____	_____	_____	_____	clean
3. quiet	_____	_____	_____	_____	_____	noizy
C. Travel/tourism						
The motorcoach was:						
1. clean	_____	_____	_____	_____	_____	dirty
2. tiring	_____	_____	_____	_____	_____	comfortable
3. quiet	_____	_____	_____	_____	_____	noizy

Thurstone scales. Thurstone scales are time-consuming to construct because they require two development stages. During the first stage, a lot of items related to the attitude being studied are submitted to a panel of judges who place each item into one of 11 categories they consider to be at equal intervals from extremely unfavorable to favorable. The final scale contains perhaps 10-20 questions whose values are equally spaced according to favorability. The final scale is thought to be more or less as interval. Clients are then asked to check statements that they disagree or agree with, and an average value for the items checked is obtained (Thurstone, 1970). The final questions might look like Table 4.6 below.

Table 4.6. General examples of Thurstone scaling.

	Agree	Disagree
1. The motorcoach I use will always have clean rest rooms.	_____	_____
2. The The TV commercial that catches my eye is always funny.	_____	_____
3. Any park I jog in will have a paved trail.	_____	_____
4. Any tour I take will have an escort.	_____	_____
5. Any sports event I attend will sell alcoholic beverages.	_____	_____

Other construction methods. There are many other ways to design questions. Whatever the method, the entire questionnaire should be designed with data collection in mind. Using 5 point scales and 7 point scales in the same questionnaire may lead to some misrepresentation during data analysis. Mixing dichotomous data and other scales on the same instrument may confuse the respondent. It may be a good idea to categorize the questions and place similar formats in the same position on the questionnaire. An example of some other question formats follows.

Table 4.7. Other question formats.

Dichotomous - A question with only two choices.	No	Yes
Did you purchase your cruise from a travel agency?		
Were the stadium seats comfortable?		
	False	True
The fitness room has adequate ventilation.		

Multiple Choice - Choose the best answer.
I prefer the following aerobic exercise:
    a. aerobic dance.
    b. bench step.
    c. Nordic Track.
    d. stair climber.
    e. treadmill.

Rank Order - Rank preference with 1 the highest, 5 the lowest.	
a. aerobic dance.	1.
b. bench step.	2.
c. Nordic Track.	3.
d. stair climber.	4.
e. treadmill.	5.

Scientifically designed surveys are difficult, time-consuming and often costly to construct. The researcher may be able to find survey examples that can be used or modified with permission for different purposes. State or federal tourism departments may have ideas for data collection in travel/tourism or sport management. City/state/federal recreation departments may have questionnaires that are usable for leisure/recreation surveys. Professional journals, theses and dissertations are other sources to explore for possible questionnaires already developed.

Interview

n interview is done in the very same manner as a questionnaire. The difference is in the way in which the data is collected. An interview instrument (questionnaire) is developed and either a personal or telephone interview is made. The interviewer should be trained to be client oriented and friendly, read the questions to the respondent, and not influence the answers with body language or different directions, unstandardized answers to meanings of words or questions asked by respondents. The interviewer will also need to be aware of sampling procedures unless assigned to specific individuals.

Normative Survey

The normative survey simply collects data from a representative sample of clients and determines "norms." The norm is found using percentiles and percentile ranks as discussed later in the chapters on data analysis. Of vital importance are the sampling procedures used in data collection. If the sample is appropriate, the data can be generalized to the population for which it was designed.

Correlation Survey

Correlation surveys do not establish cause and effect. They simply seek to determine if a relationship exists between two sets of data. For example, is there any relationship between distance between seats and perceived passenger comfort on a motorcoach? Or does a relationship exist between number of clients enrolled and number of programs in a fitness center? Although relationships could exist, there is no way to know if one factor causes another, because all factors are not controlled. Nevertheless, such a study may be used by managers to help make administrative decisions. Correlational techniques are discussed in later chapters.

Chapter 5 will discuss experimental design techniques.

Exercises

1.   List the steps in designing a questionnaire.

2.   Use your experience to design a one page questionnaire. Explain how you could establish validity.

3.   Gather data from 31 people and establish reliability by determining a percentage of agreement.

4.   Explain how to establish objectivity in a questionnaire.

5.   Define the following terms:

% of agreement	reliability	objectivity
group bias	construct validity	assumptions
occupancy rate	limitations	sanction
panel of experts	validity	pilot study

6.   Determine the percentages of agreement for each question and the overall percentage of agreement for all questions for the following data:

Table 4.8. Personal Data Percentages of Agreement.

Question #	Number Answering First Time	Number Answering Retest Same	Percentage of Agreement
03	31	31	
04	31	31	
05	29	29	
06	31	31	
07	31	25	
08	31	28	
09	31	29	
10	31	30	
11	17	17	
12	31	31	
13[1]	-	-	
14	03	03	
15	30	29	
Total	327	314	

[1]No response to this question.

References

Corbin, Dan. (1970). Recreation leadership. Englewood Cliffs, NJ: Prentice-Hall, Inc.

Latham, John. (1991). Bias due to group size in surveys. Journal of travel research, XXIX (4), 32-35.

Jubenville, Alan. (1976). Outdoor recreation planning. Philadelphia: W.B. Saunders Company.

Likert, R. (1970). A technique for the measurement of attitudes. In G.F. Summers (Ed.), Attitude Measurement (pp. 149-158). Chicago: Rand McNally & Company.

Osgood, J.C.; Suci, G. & Tannenbaum, P. (1957). The measurement of meaning. Urbana: The University of Illinois Press.

Seigel, Sidney. (1956). Nonparametric statistics for the behavioral sciences (p. 26). New York: Mc-Graw-Hill Book Company.

Thurstone, L.L. (1970). Attitudes can be measured. In G.F. Summers (Ed.), Attitude Measurement (pp. 127-141). Chicago: Rand McNally & Company.

Chapter 5

Experimental Design

Leisure, recreation, sport management and travel/tourism do not use experiments as much as other disciplines. However, a discussion of general experimental concepts may help the reader develop ideas for experimentation in these disciplines.

Experiments usually take a *control* group of persons, who serve as a comparison group, and one or more experimental groups, who have some variable(s) altered during the experiment. If known factors can be controlled or accounted for, either through manipulation or randomization, effects that emerge different from the control group are most probably due to the treatments involved. In any case, there is no definite proof, but statements can be made as to the probability being very high that differences are due to treatments, usually set at 95% or 99% probability by most researchers. Instruments, whether chemical, electronic, mechanical or written, need to be valid if the measures obtained from them are to reveal what the experiment explored.

Validity

alidity is a concept that asks if the instruments measure what they are supposed to measure. There are many different kinds of validity. Chapter 4 discussed construct or face validity in questionnaire design. The same concept also applies to instruments used in experiments. Other kinds of validity are content and criterion validity.

Construct Validity. Construct or face validity refers to how well the instrument measures the underlying theory of the factor being assessed. For example, Thurstone scales are said to measure attitudes clients have. A stress testing machine is said to measure oxygen uptake (max VO_2) when the subject exercises to exhaustion. Measured amounts of orthotolodine can determine how much chlorine there is in swimming pool water.

Content Validity. Content validity is associated with written assessments and refers to how well the instrument samples the material in question. Most knowledge tests are assumed to have this trait.

Criterion Validity. Criterion validity means the instrument is validated by comparing it to some known criterion or standard. There are two main criterion models, concurrent and predictive validly. *Concurrent validity* has a high relationship to an instrument that is already known to be valid. For example, percent of body fat determined from skinfolds measurements of selected body parts shows a high relationship to percent of body fat determined from an underwater weighing tank.

Predictive validity refers to the ability of a criterion to predict future direction. Generally, a statistical procedure - regression - is able to combine two or more factors that can be used to predict future trends. Regression techniques are discussed in a later chapter.

Aside from instrumentation, validity is also a major part of the experiment itself. Thomas and Nelson (1990) describe two main sources of invalidity in the experimental project. They are internal and external validity. Internal validity refers to whether or not the *treatments* make a difference in the experiment. External validity is concerned with the generalizability of the data. Threats to internal validity include history, maturation, testing, instrumentation, statistical regression, selection bias, experimental mortality, selection-maturation interaction, and others. Threats to external validity include reactive effects of testing, interaction of selection bias and external treatment, reactive effects of experimental arrangements, and multiple treatment inference. The reader is referred to this excellent source for learning about the threats to internal and external validity and how they can be controlled.

Variables

 ariables are concepts that are studied or that may affect the outcome of the experiment. They include categorical, control, dependent, independent and intervening variables.

Categorical. Categorical variables are described as ways that variables are classified. Examples are ethnic group, gender (male, female), religious group, skill level, etc.

Control. Control variables are those conditions that are held constant. As such, they may be described as a restricted characteristic; that is, the effect of this variable in regards to the dependent and independent variable are not studied. Therefore, the factor is held constant, or it could be randomized in order to account for it. Examples may be limiting the research to females only, or blacks only; or randomizing the clients or subjects to include both males and females, or using all cultural groups in a random fashion in the selection process.

Dependent. The dependent variable is the effect, outcome or response that occurs during the experiment. That means it is the effect of the independent variable. Examples may be decreased social distance after participating in a tour or sporting event, increased net income from a regional marketing technique, or increased fitness level from new aerobic routines.

Independent. An independent variable is one that causes the effect, outcome or response. It acts *independently* of other variables. As such, it is often referred to as the treatment effect. Examples may be a tour, if it was the tour during which decreased social distance took place; the marketing technique, if that is what caused an increase in net income; a new aerobic routine, if that is what caused an increase in fitness, etc.

Intervening. These variables are uncontrolled and act outside the bounds of the experiment. They act between the dependent and independent variables, and may be unknown or inferred during the conclusions stage. Examples may be that relief from tension and stress may have influenced social distance in a positive way during a tour or sporting event; or increased net income may have resulted from a increase in oil price, rather than the marketing technique; or improved fitness may be a result of nutrition rather than new routines. Intervening variables may bias the result of the experiment and mislead the researcher into false interpretations regarding the independent variable.

A simple, fictitious marketing experiment may help to further identify these variables. Assume that a researcher wants to know which of two colors of a brochure is most attractive to business women. Two identical brochures are made, each with the colors in question. The researcher decided to use a biased sample of business women recruited in an airport. No control group was used in this study.

A room was rented in the airport lobby, and women interviewers were trained to randomly approach every 10th. person they thought appeared to be a business woman. The person was asked to take part in the experiment, which would only take about two minutes of their time. It should be apparent that most conditions can be controlled in this situation.

The experiment consisted of asking each person identified as a business woman who agreed to participate in the study which of two brochures they liked the best. The only difference in the brochures was the color. The position of the brochures was rotated with each person. Thirty one women were interviewed and the results compiled.

Although there are many flaws with the over-all design, try to identify the variables just discussed. The categorical variable was the **business women** interviewed (classification). In this case, the categorical variable was also the control variable, because *business women* were the *only* subjects studied. The dependent variable was the *number of responses* or *choice* about color. The brochure *color* caused the effect of choice, and as such is the independent variable. There were probably many intervening variables, such as color blindness; time before next flight; whether outbound or inbound; food, drink or restroom necessity; type of business the person was engaged in; managerial level of respondents; only airport travelers interviewed; etc. Any of the aforementioned and many more variables may have influenced the choice of color. It would certainly be difficult to generalize results to the total population of business women, but it may help in identifying variables.

Design

The experimental design can take many forms. Of critical importance is a consideration of how the data will be analyzed, because some statistical techniques predicate the number of clients or subjects as well as the randomization process. Also, selecting the type of statistical analysis prior to starting the research is vital, because it is imperative to use the most powerful analysis that meets the data assumptions. In chapters that follow, assumptions of many statistical techniques are listed. Sometimes the assumptions can be violated with little effect on the outcome of the analysis, but often using a statistical process that does not meet the assumptions of the data invalidates the interpretations.

After determining instruments are valid, reliable and objective (chapter 4) and after identifying variables in the project, the design of the experiment can be determined. Then design ways the subjects will be selected and observed (tested). Steps to use in experimental design research are found in chapter one and will not be repeated here.

Types of designs include pre-experimental, experimental and quasi-experimental (Campbell & Stanley, 1963). Symbols in the tables that follow include Tr (treatment or independent variable), O (observation or test; O_1 = first test, O_2 = test 2, etc.), R (randomization of group). The statistical techniques described can be found in the chapters on data analysis.

Pre-experimental. Pre-experimental designs control few threats to internal and external validity. **Table 5.1** describes these designs.

Table 5.1. Pre-experimental designs.

	Design	Statistical Techniques	Uncontrolled Threats
1	Tr　O	Elementary	Everything
2	O_1　Tr　O_2	t (repeated measures)	History Maturation Testing
3	Tr　O_1 　　O_2	t (independent)	Selection bias Selection maturity

Experimental. These designs are experimental because groups are randomly selected. **Table 5.2** on the following page summarizes these designs.

Table 5.2. Experimental designs.

Design	Statistical Techniques	Uncontrolled Threats
4 R Tr O_1 R $\quad$ O_2	t (independent)	History Instrumentation Experimental mortality
R Tr$_1$ O_1 R Tr$_2$ O_2 R $\quad$ O_3	Simple ANOVA Factorial ANOVA Multiple ANOVA	
5 R O_1 Tr O_2 R O_3 $\quad$ O_4	Simple ANOVA Factorial ANOVA ANCOVA t (dependent) (2 X 2)	Testing (Reactive- interactive effects)
6 Solomon Four Group Design: R O_1 Tr O_2 R O_3 $\quad$ O_4 R $\quad$ Tr O_5 R $\quad\quad$ O_6		

<u>Quasi-experimental</u>. Quasi-experimental designs control most, but not all threats to internal validity. **Table 5.3** summarizes these designs.

Table 5.3. Quasi-experimental designs.

	Design
7	Time Series O_1 O_2 O_3 O_4 Tr O_5 O_6 O_7 O_8
8	Reversal design O_1 O_2 Tr$_1$ O_3 O_4 Tr$_2$ O_5 O_6
9	Nonequivalent control group: O_1 Tr O_2 O_3 $\quad$ O_4
10	Ex Post Facto Use design 3 to investigate factors statistically

Part 2 will begin data analysis using grouped data. Chapter 6 discusses characteristics of data.

Exercises

1. Visit the library and obtain a copy of the reference to Thomas & Nelson (1990). Explain the threats to internal and external validity.

2. Define the following terms:

categorical variable	criterion validity	variable
control variable	predictive validity	validity
construct validity	concurrent validity	content validity
dependent variable	independent variable	instrument
intervening variable	pre-experimental	experimental
quasi-experimental		

References

Campbell, Donald T. & Stanley, Julian C. (1963). Experimental and quasi-expermental designs for research. Copyright © 1963. Adapted by permission of Houghton Mifflin Company.

Thomas, Jerry R. & Nelson, Jack K. (1990). Research methods in physical activity (pp. 297-306). Champaign: Human Kinetics Books.

Part 2

Grouped Data

Chapter 6

Characteristics of Data

1. Definitions

Data is a collection of values or magnitudes of a quantity called a variable. A *variable* is something that can be measured and could be any of a number of values. A *population* are all the values of a measure that could be obtained. For example, if all the students enrolled in school at a given time is 2,000, whatever measures are recorded must include all 2,000 students in order to be population data. Most data are not population scores, but contain only a portion of the total that is selected to represent a population. These types of data are known as samples.

2. Kinds of Data

Numbers may be classified as continuous or discrete. *Continuous* data has an unlimited number of units, such as time, height, weight, length etc. The most common way for leisure services students to measure time is with a hand-held stopwatch that records time in hundredths of a second. However, automatic timing devices can internally measure time in thousandths of a second. It is obvious that devices can be made to measure time even further, so that there are an unlimited number of units that could be designed to measure time. By the same token, height, weight, length, etc. can be thought of as having an unlimited number of units, depending upon the sophistication of the measuring device.

Discrete data are exact or countable, such as the number of people enrolled, attendance, number of programs, etc. There can be no misinterpretation of the number of these kinds of units. Frequency counts are in the form of discrete data.

3. Data Scales

The kind of statistical technique used may depend on how the data is classified. From the lowest to highest classification, data is either nominal, ordinal, interval or ratio.

Nominal. When numbers can be used to *classify things*, the data are said to be nominal. Examples of nominal data include things that are identifiable only by groups, such as:

Nominal Data

- females = 1, males = 2.
- right = 1, wrong = 2.
- basketball = 1, football = 2, track = 3, etc.
- aggressive = 1, dogmatic = 2, outgoing = 3, etc.
- left = 1, right = 2.
- isolate = 1, leader = 2.
- no = 1, yes = 2.
- equipment categories.
- etc.

In the examples cited, *there is no way to determine which of the classifications are of higher or lower value* than the other. Frequency counts can be tabulated, and the number of responses in each category determined, but there is no way to know if right has a *higher* value than wrong, or that left has a *lower* value than right, or that yes has a *higher* value than no, etc. Notice that words are converted to numbers in these kinds of data. The word meanings may have no value at all, but when numbers are assigned to represent the words, the numbers appear to be precise values. Since the numbers appear to be data that have equal values on at least an interval scale, it is easy to misinterpret what the numbers represent.

When data are nominal, usable statistical techniques are those that lend themselves to countable data, such as descriptive statistics (mode, frequency counts). Chi-square is a possibility, as is expansion of the binomial and a technique called the contingency coefficient. *Nominal scales require that data be analyzed with non-parametric statistics.* Non-parametric statistics make few, if any, assumptions about the population distribution or the numbers they represent. Non-parametric techniques are powerful ways to analyze data when assumptions such as normality cannot be made.

Ordinal. Ordinal data are not only used to classify things, but have the *additional characteristic of being able to distinguish between higher and lower values.* Therefore, the data are rankable. Examples include:

Ordinal Data

- social distance scale.
- points in diving, gymnastics, etc.
- judges ratings of players.
- judging performance.
- aptitude, personality tests.
- rating scales.
- perceived exertion.
- etc.

Some of these measures, such as points in diving, rating scales, aptitude tests, perceived exertion, etc. appear to have data that are more precise. Regardless of how precise the numbers seem, they do not meet criteria for a higher form of measurement because the *distances between the numbers are not the same in every case*; that is, the distances are not similar and as such are more properly used as ranks than higher forms of data. Non-parametric statistical techniques are appropriate with ordinal data.

Interval. The interval scale meets the ordinal scale requirements, and in *addition has numbers that allow for differences between positions to be measured.* These distances between positions are similar data, and as such provide for more precise measurement with their numbers. Examples of interval data include:

Interval Data

- temperature.
- fitness tests, such as Physical Best.
- body composition.
- etc.

Interval data have an *arbitrary starting point.* Even though the data keeps the same relationships between the numbers, the 0 point could have been at a different place to achieve the same result. Parametric statistical tools may be used with interval data. Parametric statistics have different assumptions about the population the numbers represent, depending on the procedure used.

Ratio. When the *0 point is known, and the numbers include ratios between them,* the data are classified as ratio. Examples include:

Ratio Data

- distance.
- time.
- mass.
- weight.
- volume.
- cholesterol counts.
- height.
- length.
- velocity.
- speed.
- etc.

Some descriptive researchers analyze questionnaire results as if the data were interpreted as ratio numbers. It should be obvious that assigning a number to a characteristic and then analyzing the numbers does not usually meet the assumption of ratio data.

Both parametric and non-parametric measures can be used with ratio data. However, parametric data may be more appropriate, because assumptions made about the data may make the analysis more meaningful.

4. Discussion

Statistics are misused even today. Selecting the appropriate statistic involves trying to use the most powerful technique that will meet the assumptions underlying the data. Sometimes data assumptions can be violated with little effect on the outcome, whereas at other times a violation negates the value of the information gained.

The prudent investigator will first look at the data and determine the classification scale. Then the data characteristics will reveal whether or not specific assumptions underlying each statistical model are met. Finally, a choice is made of the most powerful or highest level statistic that meets the data assumptions. Inferences drawn from statistical use are no better or worse than the adherence to or violation of assumptions made about the data. Assumptions behind selected statistical models appear in the chapters that discuss these techniques.

It is not easy to determine when parametric or non-parametric techniques should be used. Different statistical methods have different data assumptions and different population characteristics. Generally, if the data are nominal or ordinal, non-parametric tools are the better choice, because no assumptions are usually made about population characteristics, whereas *parametric techniques require at least a normal distribution.* Some assumptions, such as normality may be violated with little effect on the results, but it is not within the scope of this text to identify when violations of assumptions may be disregarded without affecting the results. **Table 6.1** on the following page summarizes examples of appropriate statistics for different levels.

Data may be analyzed by using scores directly, or by grouping the scores into categories. Before calculators and computers were used extensively, grouping the data was more efficient and faster than trying to use all the data directly, and the results closely approximated actual precision. Today, most of the time data can be used in it's entirety, using computers or programmed calculators for most of the work. Using all the data is as simple as typing in numbers on a keyboard.

Yet there is no way to draw graphics without grouping data. Computers with printers or plotters can produce great looking graphs, but the data still have to be grouped electronically. Also, when preparing large amounts of data for presentation in a report, one way to display the information is by grouping data. By simplifying a lot of data into several columns and rows, the information can be viewed on one page in it's entirety. This simplification makes data easier to read and may allow easier display of relationships that are difficult to visualize flipping through page after page.

Table 6.1[1]. Appropriate Statistics.

Scale	Example	Type
Nominal	Mode	Non-Parametric
	Frequency Count	
	Contingency Coefficient	
	Chi-Square	
Ordinal	Median	Non-Parametric
	Percentile	
	Spearman P	
	Freidman ANOVA	
Interval	Mean	Parametric
	Standard Deviation	
	Pearson r	
	Regression	
Ratio	All Interval	Parametric

[1] Siegel, Sidney. (1956). Nonparametric statistics, (p.30) New York: McGraw-Hill Book Company. Modified with kind permission from the publisher.

Chapter 7 will introduce the concept of data analysis starting with grouped data before using raw scores.

Exercises

1.   Categorize the data below as to continuous (**C**) or discrete (**D**):

    a.   a body temperature of 98.6 degrees F.
    b.   attendance of 1,287 people at a special event.
    c.   a distance of 60'6" from home to the pitcher's plate.
    d.   18 of 25 programs meet certain criteria.
    e.   a wheelchair turning radius of five feet.
    f.   an archery target score of 150 points.
    g.   a bowling score of 150.
    h.   a golf score of 83.
    i .   a body weight of 105 pounds.
    j.   a ldistance of 6'6" between boxwood plantings.
    k.   16 potholes filled in 2 hours and 37 minutes.
    l.   3 hours and 40 minutes logged on a computer.

2.   Describe the following data as nominal (**N**), ordinal (**O**), interval (**I**) or ratio (**R**).

    a.   strongly disagree = 1, disagree = 2, no opinion = 3, agree = 4, strongly agree = 5.
    b.   50 hours and 20 minutes worked by a crew cheif in 1 week.
    c.   37.29 points on a difficult dive.
    d.   60 feet between bases.
    e.   18 of 25 workshop participants certified.
    f.   a travel distance of 1881 nautical miles.
    g.   the number pi (3.1416).
    h.   a body temperature of 100.2 degrees F.
    i.   net income of $230,000.00.
    j.   taxes paid of $173,000.
    k.   a rank of 3 of 10 supervisors.
    l.   jog = 1, run = 2, sprint = 3.

3.   Matching. Match the statistical tool on the right to the *beginning level* for the data on the left.

	Type Data		Statistical Tool
1.	Nominal	a.	mean.
		b.	Pearson r.
2.	Ordinal	c.	median.
		d.	regression.
3.	Interval	e.	Chi-square.
		f.	Friedman ANOVA.
4.	Ratio	g.	percentile.
		h.	mode.

4.   Are the following statements true (T) or false (F)?

a.	$5>3$.	i.	$43.99 = 44$.
b.	$2>1$.	j.	$15+(-3) = 18$.
c.	$4<3$.	k.	$-12+(-16)+4=0$.
d.	$5 = 5.00$.	l.	$5\div3=1.67$ (rounded).
e.	$-5>-2$.	m.	$4^2>8$.
f.	$-16<-14$.	n.	$-15+(-5)=-20$.
g.	$-3>0$.	o.	$-15+5=-20$.
h.	$(6+3)=(5+4)$.	p.	$-6+3=3$.

Chapter 7

Grouping Data

Most of the time, data can be used directly with the actual scores obtained. Sometimes, however, the data needs to be organized into catagories called *class limits* or *intervals*. These groupings are needed when bar graphs called histograms, curves, polygons and other graphics are constructed. Computer programs can organize this data automatically and quickly, but what can be done if such data needs to be drawn by hand? A basic knowledge of how to group data will help increase application of graphic procedures and serve as an introduction to elementary statistics.

Definitions. Before discussing grouping data, it is necessary to understand some terms. The *range* is found by subtracting the low score in the whole distribution from the high score. The range will help in determining the interval needed in each row. An example of range is:

15 18 20 22 29 33 34 38 43 55 57 59 60 62 **65**

Since the high score is 65 and the low score is 15, the range in the above example is 65-15 = 50.

A *row* is a horizontal line that tabulates all the numbers that fall between the range described by class limits in the row. *Class limits (CL)* are the high and low numbers that represent the data in the **row**. *Real limits* usually start with .5 of a number for the lowest value and end with .4 of a number for the high value. *Frequency counts* refer to the tabulation of numbers that fall on or between class limits defined for a row. A *tally* means to count and record all the numbers that belong to each row. The *interval* (*i*) is the numerical difference between lower limits of adjacent rows. It is found by dividing the overall range by the number of rows desired.

One rule to help determine the interval is to arbitrarily keep the number of rows between 10 and 20. When the number of rows approaches 10, the number of frequency counts in a row tend to be large, and when the number of rows is 20, the number of frequency counts in a row tends to become small. An ideal number of rows is 15, according to most texts. It is usually better to *make the interval an odd number* to keep the mid-point of the interval a whole number.

1. Example

iven the following attendance raw scores. Group the data into a frequency distribution with a tally.

Table 7.1. Attendance Scores From Metroplis Playground.

66	59	63	63	54	68	50	55	67
49	67	64	62	41	45	65	56	91
63	61	47	71	67	67	61	55	57
56	58	62	67	46	55	43	64	33
49	71	69	63	62	55	62	42	59
62	76	71	43	66	91	67	74	63
Σ345	392	376	369	336	381	348	346	370

Step 1. Determine the range. The range is: 91-33 = 58.

Step 2. Determine the interval. The true Interval is:

$$i = \frac{range}{desired \ \# \ of \ rows}; \ i = \frac{58}{15} = 3.87$$

Discussion. In order to have 15 rows, the interval would need to be a decimal, 3.87. This number is not practical. The interval needs to be an odd whole number, so it would appear the closest odd number to 3.87 is 3. If an interval of 3 is used, it will require 58÷3 or 19+ rows. An interval of 5 would require 58÷5 or 11+ rows.

It was decided to use an interval of 5, although an interval of 3 would fit the criteria just as well. If an interval of 3 were used, the total number of rows could go over the 20 limit, whereas if an interval of 5 is used, any additional rows over 11 moves in the desired direction towards 15 rows. This decision is a matter of judgement.

Step 3. Construct a frequency distribution tally.

Discussion. It would make little difference if the top or bottom score in the distribution were used as the starting point. This text will *always start with the high score and work down.* Take the highest score in the distribution and make it the mid-point of the top interval. Then establish limits on either side of the mid-point. An example is given in **Table 7.2**.

Table 7.2. Frequency Distribution.

Row	Class Limits (CL)	Real Limits	Tally	Frequency (f)
13	89 91 93	88.5 - 93.4	11	2
12	84 88	83.5 - 88.4		0
11	79 83	78.5 - 83.4		0
10	74 78	73.5 - 78.4	11	2
9	69 73	68.6 - 73.4	1111	4
8	64 68	63.5 - 68.4	1111 1111 11	12
7	59 63	58.5 - 63.4	1111 1111 1111	14
6	54 58	53.5 - 58.4	1111 1111	9
5	49 53	48.5 - 53.4	111	3
4	44 48	43.5 - 48.4	111	3
3	39 43	38.5 - 43.4	1111	4
2	34 38	33.5 - 38.4		0
1	29 33	28.5 - 33.4	1	1
			Σ	54

In practice, rows are rarely numbered, and row numbers will be eliminated in future distributions. Also, abbreviations for class limits (CL) and frequency (f) will be used. Finally, real limits will not be recorded. It is understood that the real limits are decimals, starting with .5 and ending with .49999 (not rounded) for tabulation purposes.

It is important to get the tally correct. Since misaligned and hard to read numbers create more chances for error, try using graph paper. Graph paper makes it easy to properly align data in rows and columns. Also, mark off each number as it is recorded. Finally, check the work a second time to insure a proper tally.

Once the data are recorded, the scores lose their exact value. It is assumed each of the frequencies are distributed normally around the mid-point of each row so that variances between scores are equal (*homoscedasity*). For example, row 6 is assumed to look like Figure 7.1 below:

```
53.5 . . . . 56 . . . . 58.4
 f  1  1  1  1  1  1  1  1  1
```

Figure 7.1. Row 6 homoscedasity.

A frequency distribution chart is one way to make data more meaningful. By using just 10 - 20 rows and several columns, a lot of data can be compiled into a short space. It makes the data more understandable than trying to see relationships in a long, unorganized list of numbers. Since the amount of data should increase in the middle rows, and taper at the top and bottom, it is easy to recognize data that does not fall into a usual, predictable pattern. Recognizing this misalignment of numbers is one way that data should begin to pose questions for further study.

Be aware that computer programs can speed up accuracy and allow tables and graphics to be integrated into one document. The major task is to accurately type numbers into rows and columns. Also, formulas can be inserted into cells, allowing for quick and accurate determination of results. Mastering any of the major spreadsheet or dara management programs will save time in the long run. A desktop computer with lots of memory is all that is needed.

Table 7.3. Attendance at Metroplis Playground.

(1) Class Limit (CL)	(2) frequency (f)	(3) Rel. f (f/N)	(4) % f (X 100)	(5) Cum. f
89 - 93	2	.037	3.70	54
84 - 88	0	.000	0.00	52
79 - 83	0	.000	0.00	52
74 - 78	2	.037	3.70	52
69 - 73	4	.074	7.41	50
64 - 68	12	.222	22.22	46
59 - 63	14	.259	25.93	34
54 - 58	9	.167	16.67	20
49 - 53	3	.056	5.56	11
44 - 48	3	.056	5.56	8
39 - 43	4	.074	7.41	5
34 - 38	0	.000	0.00	1
29 -33	1	.019	1.85	1
Σ	54			

Step 4. Draw a frequency distribution chart (see **Table 7.3** above).

Prior to starting, it is necessary to define some additional terms. Column 3, *relative frequency* (Rel. f), is found by dividing the row frequency by N and is recorded to the nearest thousandth of a decimal place. Column 4, percent of frequency (% f), is determined by multiplying the relative frequency by 100 and is recorded to the nearest one-hundredth (2 decimal places). Column 5, cumulative frequency (Cum. f) starts on the **bottom** row and adds each row frequency in succession **up** the column (see **Table 7.3** above).

Table 7.3. (Continued from page 89)

(6)	(7)	(8)	(9)	(10)
Cum. Rel. f (Cum.f/N)	Cum. % (PR) (X 100)	d	fd	fd²
1.000	100.00	+6	+12	72
.963	96.30	+5	0	0
.963	96.30	+4	0	0
.963	96.30	+3	+6	18
.926	92.59	+2	+8	16
.852	85.19	+1	+12	12
.630	62.96	0	0	0
.370	37.04	-1	-9	9
.204	20.37	-2	-6	12
.148	14.81	-3	-9	27
.093	9.26	-4	-16	64
.019	1.85	-5	0	0
.019	1.85	-6	-6	36
		Σ	-8	266

Column 6, *cumulative relative frequency* (Cum. Rel. f) is the cumulative frequency divided by N and taken to the nearest thousandth decimal. It can also be obtained by adding relative frequency in each successive row, starting at the bottom row and adding upwards, being careful not to add any cells that have a frequency of 0.

Column 7, cumulative percent (Cum. %), is also known *as percentile rank* (PR), and is the cumulative relative frequency multiplied by 100. Cum % is recorded to the nearest hundredth. Examples of this data are found in **Table 7.3** on the previous page.

Other columns are d, fd, and fd^2. The d column (8) is located by finding the *assumed mean* (AM). The assumed mean is the middle score in the distribution. It is found by $N \div 2$. Once the position of the AM is known, the AM **score** is assumed to be the **midpoint** of the interval where that score lies. Starting in the row that contains the AM, a 0 is recorded. Working up the column to the topmost rows, simply add 1 to each row, recording the + signs. From 0 down, add minus 1 to each row, recording the - signs. The algebragic sum of this d column is needed to find the mean score. An example is shown in **Table 7.3** on page 90.

The fd column (9) is the product of the f and the d columns in each row. The fd^2 column (10) is the product of the d and fd columns in each row. The sum of the fd^2 column is needed to find the standard deviation. Examples of fd and fd^2 are shown in **Table 7.3**.

To summarize, the following review may be helpful in constructing a frequency distribution chart:

1   Columns are cumulative vertical blocks, while *rows* are horizontal spaces. Eleven columns are needed for a chart (if the tally is included. If a tally is not desired, only 10 columns are used). If standard scores are desired, the chart will need 18 columns (17 without the tally column).

2.   Determine the class limits.

3.   Arrange the data in a convenient manner and make a frequency count of each number. Mark off each score as the count is made. If the tally (count) is not correct, related statistics will be inaccurate.

4.   Enter the numerical tally totals for each row in the f column. Sum the column and check to see that the total frequency agrees with the total number (N) of scores.

5.   Determine relative frequency by dividing f by N. Record the relative frequency for each row to the nearest thousandth (.000).

6.   Calculate the percent of frequency by multiplying the relative frequency by 100, and record to the nearest hundredth (0.00).

7.   Determine the cumulative frequency by adding f in each row, starting at the *bottom* and *adding upward*. Check to see the total at the top agrees with N or the total of the f column.

8.   Find the cumulative relative frequency by adding the relative frequency column from the bottom up. Another way is to divide the cumulative frequency column by N. Record this number to the nearest thousdandth (.000).

9.   Record the cumulative percent column (percentile rank) by multiplying the cumulative relative frequency column by 100, and record to the nearest hundredth (0.00).

10.   Compute the assumed mean. The assumed mean is what the mean is estimated to be. If all scores were arranged from high to low, and the data is normally distributed, the assumed mean should be the middle score.

In a frequency distribution, the data have been grouped, and there is no way to know the position of the true mean. But it can be assumed the data are normally distributed, and the mean score should be in the row where the Cum. f that represents half of the scores lie.

How To Find The Assumed Mean (AM):

a.   Start with the cumulative frequency column. Visualize this column as an array with **every** original score listed from top to bottom, starting with score number 54. In **Table 7.3**, The halfway point should be half of N or position number 27 (54÷2 = 27). There is no position of 27 listed in the table, but there is a cumulative frequency of 20 and the next higher cumulative frequency is 34. A position of 27 must lie between 20 and 34. Therefore, a cumulative frequency of 27 will be above the row containing a position of 20, but below 34. That means a score corresponding to a position of 27 must be in the row with class limits of 59-63.

b.   Find the midpoint of the class limits where the assumed mean lies. The midpoint of the limits just found (59-63) is 61. Therefore, the assumed mean is a score of 61.

11.   Construct the d (difference) column by placing a 0 in the row where the assumed mean lies. Starting at 0, add 1 to every row in the d column from 0 to the top. From 0 down, subtract 1 in every row. Although this procedure is mechanically correct, what is being done is dividing the midpoint of every interval by the assumed mean.

12.   Determine the fd column by multiplying the f column times the d column. Be sure to post the proper sign (+ or -). Add algebraically and record the sum at the bottom of the column. A plus, minus or 0 number can occur. Although the analogy is not entirely correct, it is like multiplying the value of scores times the number scores with that value. Just as an average score is found by adding all the scores and dividing by N, these values will later be added and divided by N (after considering the interval width) to get the mean score.

13.   Determine the fd^2 column by multiplying the fd column by the d column. Add the total and record the sum at the bottom. Later, when sums of deviations squared, variance, and standard deviation are found, the sum of this column will be used. It is similar to adding column totals of raw scores squared.

The d, fd and fd^2 columns are used to determine the mean and standard deviation. Although grouped data results are not exact, these numbers are usually very close to the true raw score values. However, computer technology generally invalidates use of a frequency tally to determine these and similar measures. But the frequency distribution chart is still useful when arranging data for inspection, when drawing graphics and when displaying large amounts of data on one page.

2. Graphics

Once a frequency distribution has been prepared, graphic representations of the data can be drawn. Graph paper is needed, and it is customary to plan convenient units before constructing the graph. The X or horizontal axis is known as the **abscissa**, while the Y or vertical axis is known as the **ordinate.** For uniformity and comparison purposes, the ordinate height in this text will be constructed at approximately ⅔ of the abscissa length. To obtain the ordinate height, multiply the number of units in the abscissa by .67 and use that figure for the length of the ordinate. Trial and error will result in a convenient number of units for both axes.

One type of graph is a histogram. A *histogram* or *bar* graph is a plot of frequency against class limits. To construct a histogram (see Figure 7.2), use the following procedure.

1.   Plot the **score** or class limit values on the X (horizontal) axis. First, calculate a convenient number of units on the graph to use and plot the limits. Be sure to label the axis as class limits or identify the scores.

2.   Plot frequency values on the Y (vertical) axis. Make the height approximately ⅔ of the base (60-75% is acceptable). Example: height = 23 units, base = 32 units. 23÷32 = 72%. It takes practice, but most graphs can can be constructed very close to the ⅔ porportion either manually or in a computer print-out. Label the Y axis as frequency (f).

It should be pointed out that scores plotted on the X axis could begin anywhere on the X line, but the data for the Y axis must originate on the X axis. This procedure is necessary so that different graphs can be compared with consistency. The student should begin to mentally picture areas below plots as representing percentages of areas relating to the data.

Color and/or shading can help to make a histogram more understandable. A graph of the data is usually easier to understand than a table. Comparing one table of values with another is tedious and time-consuming, particularly when numbers are not rounded to simple whole units, but a quick glance at the same data in graphic form is usually easily understood.

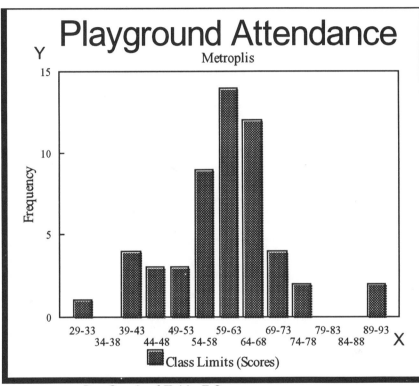

Figure 7.2. Bar Graph of Table 7.3.

This text will assume that all graphics are drawn by hand. Yet the same procedures apply when using data management computer programs. Program techniques may be different, but an understanding of how to draw graphics manually helps in utilizing programs more efficiently. Since different programs require different techniques, no generic method is adequate to describe how to create computer graphics. However, the person who can master the art of drawing graphics by hand can include these same ideas when using computers. A basic knowledge of graph axes along with understanding which data in a column of numbers belong to which axis is vital to create computer graphs. Figure 7.2 above is a bar graph of frequency from **Table 7.3**.

To interpret Figure 7.2, notice the data generally appear to follow a bell shaped curve. The majority of the scores cluster around class limits of 59-63, but there are extreme scores on either end (29 -33 and 89-93). Gaps in the limits 34-38, 79-83 and 84-88 stand out readily. These gaps ask "why were there no scores within these ranges?" The data do imply general normality, although everyone might not agree with this interpretation.

A *frequency polygon* is a line graph of the frequency. To construct a frequency polygon, follow the same method as in a histogram, except midpoints of the CL are plotted instead of lower limits. The polygon is a straight line connecting the midpoints (see Figure 7.3 below).

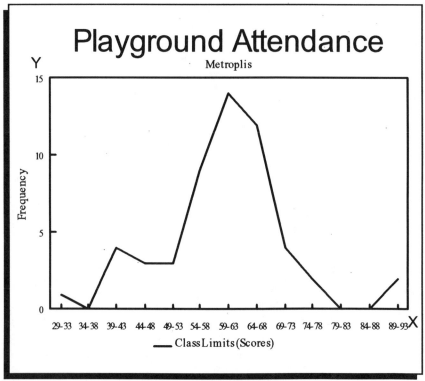

Figure 7.3. Frequency Polygon of Table 7.3.

In interpreting the polygon, notice that the frequency of 0 in the 34-38 range does not stand out as clearly as the gap in the same CL in the histogram in Figure 7.2. Yet the general normality of the data is more apparent than the same data shown by the bar graph. Also, a new gap in the CL 44-48 is more visible than it was in the histogram.

In a similar fashion, look for gaps in data of all graphs to help ask questions about what the data mean. The data *are not answers* to questions; rather, data should ask questions that motivate the mind to find answers about why things happened.

To construct a *frequency curve*, follow the same procedure as for a polygon, except round off any sharp corners of the polygon. The smoother appearance is often more pleasing to the eye. With computers, the rounded line can sometimes be drawn manually. No example of a curve will be given.

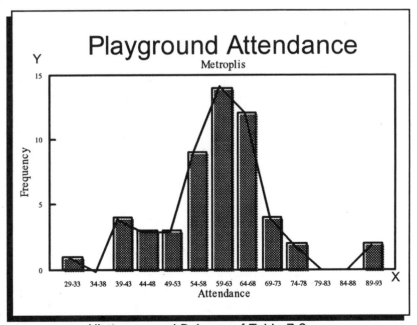

Figure 7.4. Histogram and Polygon of Table 7.3.

A histogram, polygon and/or curve can be drawn on the same graph. Figure 7.4 above is an example of a such a graph.

An *ogive curve* is a plot of cumulative frequency (Cum. f) or percentile rank (PR). This curve is used to predict PR from a score or vice-versa. Although computers can quickly tabulate PR, finding data in a long line of scores or in a large table can be time-consuming. The ogive curve has data condensed in a readable graph that helps to make a lot of data more understandable. To construct an ogive curve, use the following procedure (see **Figure 7.5**).

1. Plot score values on the X axis. Determine a convenient number of units. Plot and label this axis with units of measure the data represents, such as weight, HDL, etc.

2. Plot and label cumulative percent on the right Y axis. One way to determine the units is to make the top value equal to 100%, then divide the rest of the distance into combinations of deciles (10, 20, 30, etc.). An alternate way to determine units is to make the top value equal 100%, then mark half way down as 50%, ¾ way down as 25%. etc. Plot other points as convenient.

3. Plot and label cumulative frequency on the left Y axis. Since the cumulative frequency data is on the same row as the percentile rank data, it may be possible to identify cumulative frequency data that corresponds to PR. For example, matching a Cum. f of 20 with a PR of 37% makes the Cum. f plot easier to locate on the left Y axis. By the same token, it may sometimes be easier to *start* with the cumulative frequency side.

4. When the X and Y axes are complete, plot each of the class limits on the graph against cumulative percent for the same row from the table. Connect each plotted point with a line.

Notice that the cumulative frequency and cumulative percent points coincide on the graph. The graph can be used to predict percentile rank (cumulative percent) for any score (class limit). To do so, move perpendicular from the starting axis to the curve. Then go directly to the next axis and read the predicted data. These data are not exact values.

For example, to find the PR represented by an attendance of 59, locate 59 on the PR axis. Draw a line perpendicular to the X axis until it intersects the curve. From that position, draw a line perpendicular to the right Y axis. Where this line crosses the Y axis, read the percentile rank from the PR column (Y axis). In the example given, an attendance of 59 represents a PR of about 60%.

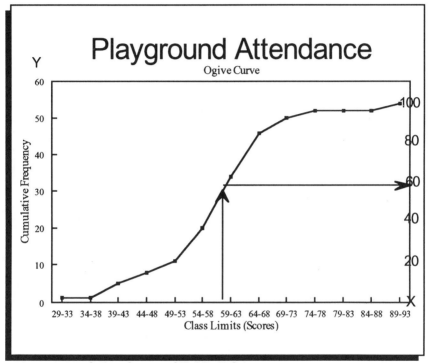

Figure 7.5. Ogive Curve of Table 7.3.

A *pie* chart is useful to present generaizations. Computer generated pie graphs can often copy labels and data from a table and incorporate these into the picture. A popular way to display this type of data is with percentages (f/N).

To draw a pie graph, first determine the percent of each data plot. Then multiply the percentage by the number of degrees in a circle, 360. The resulting number of degrees is measured by a protractor and plotted on the graph.

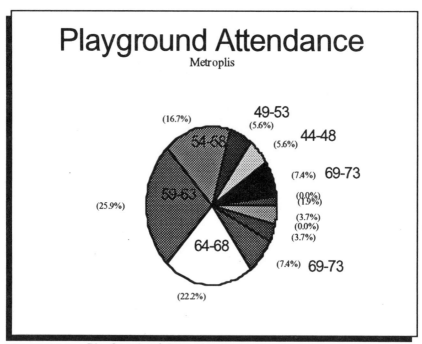

Figure 7.6. Pie Graph of Attendance.

It is also possible to use other kinds of graphs, such as a combination of a bar with a *line* above it. The line may indicate a trend of one variable, while the bar represents a different variable. Gaps in a mixed graph as described are easy to identify.

Various combinations of lines, bars and curves can be used. However, information plotted with different displays of the same data are more difficult to interpret. Different colors for each plot would make the graph easier to read. Generally, the simplier a graph is made the easier it is for the reader to interpret it.

Curves are especially useful in data analysis. A major concept when analyzing data has been a bell-shaped or normal curve. It has been found that if a lot (about 10,000) of scores are randomly obtained, the frequency curve assumes the shape of a bell. Percentages of area for various points on the X axis have been determined, and the phrase normal curve is widely used (see Figure 7.7 on the next page).

It is very important to be able to apply the concepts shown in the normal curve below. The relationships of standard deviation to z scores is very meaningful. Note the z and σ numbers are identical. Also, note the relationships between percentile and percentile rank. Memorize the percentages of positive and negative areas under the curve associated with each standard deviation. Taking time to

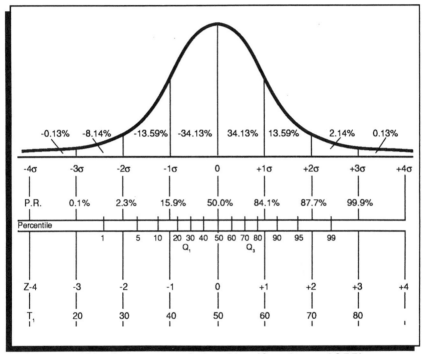

Figure 7.7. Normal curve relationships (Seashore, 1955).

recognize these simple relationships now will lead to a faster understanding of concepts based on this curve that will follow. There are other kinds of curves, as shown in Figure 7.8 on the next page.

Graphics are not the only information that can be obtained from grouped data. Measures of central tendency such as the mean, median and mode can be found, using techniques that are similar to familiar raw scores. Variance, standard deviation percentiles, percentile rank and other analyses can also be determined. Chapter 8 will explain some of the methods used to analyze grouped data.

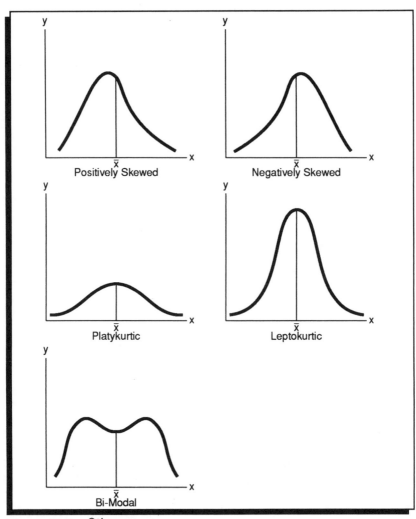

Figure 7.8. Other curves.

Applications

Management. Financial data, numbers of people, etc. lend themselves well to all kinds of graphucs. For example, years can be plotted on the X axis while sales, assets, liabilities, number of clients, attendance, etc. can be plotted on the X axis. Of particular interest is a mixed graph, where a line can be plotted above a bar graph to reveal relationships, such as net income for the bar portion and sales as the line. When the units involved are the same, such as dollars and the data scales are not far apart numerically, such a graph can be drawn.

Example: The following data shows net income and sales of American Recreation Centers, Inc. from 1984-1988 (Walsh, 1991).

Table 7.4. Net Income vs. Sales.

	Net Income	Sales
1984	1,961,000	22,313,000
1985	1,667,000	23,537,000
1986	84,000	24,818,000
1987	1,816,000	28,650,000
1988	811,000	38,356,000

Notice how Figure 7.9 reveals net income dropping from 1984-1988, even though sales almost doubled. Note the 1986 bar gap.

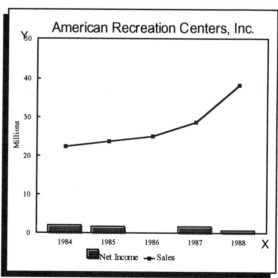

Figure 7.9. Net Income and Sales, 1984-88.

Travel/Tourism. Many examples exist in the travel/tourism field. Of particular importance is the annual Outlook for Travel and Tourism published by the U.S. Travel Data Center. In addition to graphics, tables are compiled that can be converted into graphics. Usually, the latest data for international travel, auto travel, attractions, family vacationing, air travel, accomodations, business travel, bus travel, foodservice and bed and breakfast inns are published and analyzed. An example of data that could be graphed follows (U.S. Bureau of the Census, 1990).

Table 7.5. Index of Output
Per Hour - Air Transportation

1982	115.8
1983	141.9
1984	152.6
1985	162.1
1986	178.5

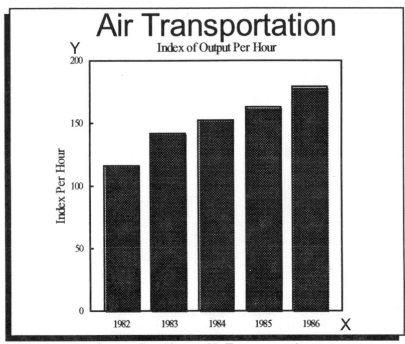

Figure 7.10. Output Per Hour - Air Transportation.

Other examples of graphs follow (U.S.Bureau of the Census, 1994).

Table 7.6. Pleasure Travel to the U.S., 1990

Country	Actual Travelers	Potential Travelers
Germany	385,000	9,968,000
U.K.	956,000	4,183,000
Mexico	1,024,000	7,802,000
Japan	1,501,000	8,772,000
Canada	2,792,000	12,844,000

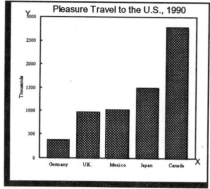

Figure 7.11. Pleasure Travel to the U.S., 1990.

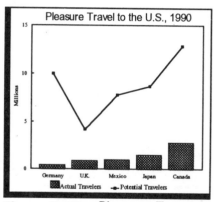

Figure 7.12. Pleasure Travel to the U.S., 1990.

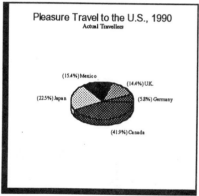

Figure 7.13. Pleasure Travel to the U.S., 1990

Chapter 7 has discussed ways to group data and create graphics. Chapter 8 will explain measures of central tendency and position using grouped data.

Exercises

1. Given the following client ages:

Table 7.7. Ages of Clients Using Metroplis Center.

38	64	50	32	44	25	49	57	46	58
40	47	36	48	52	44	68	26	38	76
63	19	54	65	46	73	42	47	35	53
40	35	61	45	35	42	50	56	45	28

    a. Complete a frequency chart with tally including CL, tally, f, Rel. f., % of f, Cum. f., Cum. Rel. f, Cum. %. Use i = 5.

    b. Construct a histogram, polygon, curve and ogive for the data.

2. One hundred outdoor sites recorded the following visits:

Table 7.8. Number of Visits to Outdoor Sites.

76	80	57	80	86	76	67	75	76	72
84	83	82	87	68	66	60	60	69	82
66	88	73	58	76	75	85	95	60	73
86	79	95	84	42	76	87	74	74	56
50	80	68	98	80	68	79	73	72	74
78	77	60	88	40	82	77	87	78	82
66	84	85	72	59	63	76	53	57	78
79	92	80	68	90	75	74	98	64	45
68	57	79	76	83	35	76	88	62	75
60	53	63	80	94	34	78	64	58	58

2. a. Complete a frequency table **using** *i* = **5** with CL, tally, f, Rel. f, % of f, Cum. f, Cum. Rel. f, Cum. %, d, fd and fd².

    b. Draw a histogram, polygon, curve and ogive for the data.

3. Given an N of 20 with a range of 45. What should the interval be?

4. The tail of a curve points to the right. It is a _____ skewed curve.

5. True or false. The abscissa and X axis are the horizontal scores, while the ordinate and Y axis are vertical data.

6. Given the following data:

Table 7.9. Fictitious Distribution.

Row	CL	f	Rel. f	% f	Cum. f	Cum. Rel. f	Cum. %
12	95-99	1					
11	90-94	2					
10	85-89	3					
9	80-84	4					
8	75-79	5					
7	70-74	8					
6	65-69	10					
5	60-64	7					
4	55-59	4					
3	50-54	3					
2	45-49	2					
1	40-44	1					

    a.   Complete the rest of the chart.

    b.   Draw a frequency polygon and ogive curve for the data.

    c.   Does the data approximate a normal curve? Explain.

    d.   What is the mid-point of row 8?

    e.   What is N?

    f.   What are the upper limits of row 4?

    g.   What are the lower limits of row 10?

7.   Given the following playground attendance figures:

Table 7.10. Playground Attendance at 36 Sites.

29	50	36	28	54	49
33	53	40	38	37	56
35	57	47	46	41	51
43	29	55	30	52	43
44	25	26	32	58	39
27	31	42	48	34	45

    a.   Determine the proper interval and construct a frequency chart, including CL, tally, f, Rel. f, % of f, Cum. f, Cum. Rel. f, PR, d, fd, and fd^2.

    b.   Draw frequency and ogive curves for the data.

    c.   Using the ogive curve, estimate $PR_{25, 33, 50}$.

8.   Given the following data (Walsh, 1991):

Table 7.11. Enchanted Parks, Inc. Data.

Year	Net Income	Sales
1985	$85,000	$3,851,000
1986	$689,000	$3,651,000
1987	$277,000	$4,684,000
1988	$22,000	$4,489,000
1989	$14,000	$4,554,000

a.   Draw a bar graph showing net income.

b.   Create a mixed graph with net income as a bar and sales as a line.

c.   Draw a pie graph for sales.

9.   Use the following data (Walsh, 1991) to draw graphics.

Table 7.12. Caesars World, Inc.

Year	Net Income	Sales
1985	$31,812,000	$654,845,000
1986	$41,017,000	$693,813,000
1987	$33,695,000	$784,740,000
1988	$77,032,000	$832,975,000
1989	$66,874,000	$902,035,000

a.   Create a bar graph for sales.

b.   Draw a mixed graph with net income as a bar and sales as a line.

c.   Draft a pie graph for net income.

10.   The following data is taken from Walsh (1991):

Table 7.13. Caesars World, Inc.

Year	Net Income	Equity
1985	$31,812,000	$226,575,000
1986	$41,017,000	$269,774,000
1987	$33,695,000	$417,441,000
1988	$77,032,000	$172,287,000
1989	$66,874,000	$243,918,000

a.   Design a mixed graph with equity as a line and net income as a bar.

b.   Sketch a pie graph of equity.

c.   Return on Equity (ROE) can be defined as Net Income ÷ Total Equity. Using this procedure, find ROE for all 5 years. Draw a pie graph of ROE.

References

Seashore, H.G. (Editor). (January, 1955). Test Service Bulletin (8). New York: The Psychological Corporation.

U.S. Bureau of the Census. (1990). Statistical Abstract of the United States, (110th. ed.) (p.405). Washington, DC: U.S. Bureau of the Census.

U.S. Department of Commerce. (1994). U.S., Industrial Outlook 1994, (p. 410). Washington, DC: U.S. Department of Commerce.

Walsh, Robert (Publisher). 1991. Walkers Manual of Western Publications, (pp. 378, 781). San Mateo, CA: Walker's Western Research.

Chapter 8

Statistical Techniques Using Grouped Data

1. The Mean ($\bar{X}$)

I n a normal distribution, the mean is usually the measure used to describe central tendency. Using all the scores, the *mean* or average *is the sum* (Σ) *of the data divided by the number of units* (N). In grouped data, the mean score is assumed to be the middle score in the distribution. However, information about the middle score needs to be adjusted because grouping the data has resulted in a loss of accuracy.

The fd column is similar to a column of raw score numbers, arranged in descending order from high to low. It has been created by subtracting the assumed mean from the mid-point of each row. Therefore, in order to convert the coded (fd) scores to their original values, it will be necessary to add the assumed mean (*AM*) back into the computations. Also, the fd column was further coded by dividing each row entry by the interval. This procedure must be compensated for by multiplying the interval back into the coded data.

Finally, the sum of the fd column is similar to the sum of a column of raw scores, so an average score must allow for division by N. The procedure below shows these steps and will result in an average score that approximates the true mean.

(1) $$\bar{X} = AM + \left[\frac{\Sigma fd}{N}\right] * i$$

Where: $\bar{X}$ = *mean*; *AM* = *assumed mean*
Σfd = *sum of fd column*
i = *interval*
N = *number of scores*

Notice how similar the procedure above is to the usual method of adding all the data and dividing by N. The Σfd is similar to the sum of all the true scores. Allowances made to add back the *AM* and *i* are the major differences in the two methods.

Row	C.L.	f	rel. f (F/N)	% f (X 100)	Cum. f	Cum. Rel. f (Cum. f/N)	Cum. % (PR) (X 100)	d	fd	fd²
13	89-93	2	0.037	3.70	54	1.000	100.00	+6	+12	72
12	84-88	0	0.000	0.00	52	0.963	96.30	+5	0	0
11	79-83	0	0.000	0.00	52	0.963	96.30	+4	0	0
10	74-78	2	0.037	3.70	52	0.963	96.30	+3	+6	18
9	69-73	4	0.074	7.41	50	0.926	92.59	+2	+8	16
8	64-68	12	0.222	22.22	46	0.852	85.19	+1	+12	12
7	59-63	14	0.259	25.93	34	0.630	62.96	0		0
6	54-58	9	0.167	16.67	20	0.370	37.04	-1	-9	9
5	49-53	3	0.056	5.56	11	0.204	20.37	-2	-6	12
4	44-48	3	0.056	5.56	8	0.148	14.81	-3	-9	27
3	39-43	4	0.074	7.41	5	0.093	9.26	-4	-16	64
2	34-38	0	0.000	0.00	1	0.019	1.85	-5	0	0
1	29-33	1	0.019	1.85	1	0.019	1.85	-6	-6	36
	Σ 54								-8	266

Figure 8.1. Data reproduced from Chapter 7.

Using the data from **Table 7.3** (Chapter 7) in Figure 8.1 reproduced above, the following example for determining the mean with procedure 1 on page 35 is shown below:

$$\bar{X} = 61 + \left[\frac{-8}{54}\right] * 5$$

$$\bar{X} = 61 + \left[\frac{-40}{54}\right]$$

$$\underline{X} = 61 + [-.74]$$

$$\bar{X} = 60.26$$

As a comparison of the mean for grouped data with the mean when all of the scores are used, the sum of the data for this distribution was 3,263. With an N of 54, the true mean is 3,263÷54, or 60.43. The mean of 60.26 obtained through group analysis is very close to the true mean.

The mean is a common statistic, but it can be greatly influenced by extreme scores at either end of a column of numbers. When scores are normally distributed, the mean is the best measure of central tendency.

2. The Mode (Mo)

The *mode is the score of greatest frequency*. In grouped data, it is the midpoint of the interval with the score of greatest frequency.

Using the data in **Table 7.3**, the row with the greatest frequency is the one that has a frequency of 14 (59-63). The midpoint of the interval is 61. Therefore, the mode *score* is 61. By contrast, the mode using raw data (all the scores) is 67.

There can be more than one mode in some distributions. If so, be sure to list **all** of them.

3. The Median (Md)

he *median is an imaginary score that has the same number of scores above and below it.* It is found by placing all of the scores in a column from high to low, top to bottom. If N is an even number, the median will lie halfway between the two middle numbers. If N is odd, the median is the middle number.

Using grouped data, there are different ways to find the median. One method is to use the same technique that percentage is determined. Some statisticians do not agree with this procedure, but it is useful in allowing the same way to find both the median and percentile scores. Equation 2 below illustrates how the median is found.

$$(2) \qquad Md = ll + \left[\frac{Number \ of \ scores \ needed}{Number \ of \ scores \ in \ row} \right] * i$$

Determining the Median. Using the data from **Table 7.3**, the following example for finding the median is given:

1. Determine the lower limits (*ll*):

    a. Calculate the *assumed median rank (AMR):*

    $AMR = (N+1) \div 2 = (54+1) \div 2 = 27.5.$

    b. Discover the row where the *AMR* is found in the Cum. f column.

    The *AMR* (27.5) is not listed. However, a Cum. f of 20

is in row 8, and row 7 has a Cum. f of 34. 27.5 must lie in row 7, because 20 is the upper Cum. f limit of row 8.

c.   Find the lower limits. Since the AMR is positioned in row 7 with CL of 59-63, **// = 58.5.**

2.   Determine the *number of scores* needed to go into the interval.

    a.   Since the *AMR* is in row 7, the CL of row 7 contains the interval desired. The *real limits* of this interval are 58.5-63.4. According to the f column, there are 14 scores in this interval. Based on the assumption of homoscedasity, the CL of this interval should look like Figure 8.2.

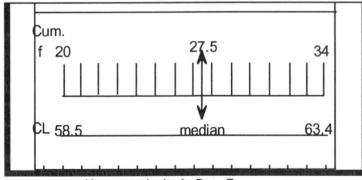

Figure 8.2. Homoscedasity in Row 7.

Where **//** =   lower limits of the row where the assumed median lies.

Number of scores needed = the number of scores to go into the row where the AM lies. It is found by subtracting Cum. f in the row below the assumed median from the computed median rank (N+1)÷2.

Number of scores in row = f in the row where the assumed median lies.

i =   interval.

Although the value of the median is not known, it is shown as an assumed rank or Cum. f of 27.5 (Figure 8.2) as determined in 1.a by $(N+1) \div 2$. It is also known that the 20th. rank or Cum. f ended at the top of row 8. Therefore, the median is 27.5 minus 20 or 7.5 scores from the start of row 7. The *number of scores needed* is 7.5.

3. Discover the *number of scores* (f) in the interval (CL) where the median is located. The **f** in row 7 **is 14.**

4. Record the interval. **i was constructed as 5.**

5. Substitute the known values into the equation below.

$$Md = ll + \left[\frac{Number\ of\ scores\ needed}{Number\ of\ scores\ in\ interval} \right] * (i)$$

(3)
$$Md = 58.5 + \left[\frac{7.5}{14} \right] * (5)$$

$$Md = 58.5 + \frac{37.5}{14}$$

$$Md = 58.5 + 2.6 = 61.18$$

If each score was listed in ascending order, low to high, the 27.5th. rank (median) would be 61.18.

The median is not influenced by extreme scores at either end of the distribution, whereas the mean score is affected by all scores that are used. The median is a way to focus more closely on central tendency when the data contains extreme scores.

The mean, mode and median are the tools used to describe central tendency, or the trend of the data towards the middle score. As mentioned in Chapter 7, a normal or bell shaped curve has the mean, mode and median as the same score. If these numbers have similar values, the distribution may approach normality. However, if these statistics are radically different, the data cannot represent a normal curve.

Other Measures

4. Percentile

 percentile is a score at or below which a given percent of cases lie. A percentile is found the same way the median was found. However, the median position was found by using (N+1)÷2. Percentile positions that are needed to determine lower limits are found by multiplying the percent in question by N. For example, using **Table 7.3** the lower limit of the 75th. percentile is found by 54 (N)*(.75) = 40.5.

The CL where the 75th. percentile lies is the row that contains the 40.5th. score, or 64-68. This row is found by moving up the cumulative frequency column until the row containing the 40.5th score is located.

To locate the proper row, an examination of the cumulative frequency column in **Table 7.3** reveals the 34th. score ended in the row with limits of 59-63. Above this row, the CL 64-68 contains the 46th. score. The 40.5th score must lie *above* the row that ends with the 34th. score, so the 40.5th. score is located between 64-68. The lower limit of the interval 64-68 is 63.5. The lower limit for all percentiles is found in the same way, using the appropriate percent figure. An example using P_{75} and P_{25} is shown below.

$$P_? = ll + \left[\frac{\textit{Number of scores needed}}{\textit{Number of scores in row}}\right] * i$$

(4)

$$P_{75} = 63.5 + \left[\frac{6.5}{12}\right] * 5 \qquad P_{25} = 53.5 + \left[\frac{2.5}{9}\right] * 5$$

$$P_{75} = 63.5 + \left[\frac{32.5}{12}\right] \qquad P_{25} = 53.5 + \left[\frac{12.5}{9}\right]$$

$$P_{75} = 63.5 + 2.71 = 66.21 \qquad P_{25} = 53.5 + 1.39 = 54.89$$

Note that a percentile is a *score*, not a percent (%). The name percentile is confusing unless this characteristic is kept in mind.

5. Quartile

uartiles (Q_i) *are convenient ways of grouping data expressed in percentiles of 25.* $Q_1 = P_{25}$, $Q_2 = P_{50}$; $Q_3 = P_{75}$. For example, students below the 25th. percentile could all be grouped together as in the first quartile. A student scoring in the first quartile would have a low score, while a student scoring above the 3rd. quartile would have a very high score. Note that $Q_2 = P_{50}$.

6. Decile

eciles are percentiles with increments of ten instead of four, as with quartiles. $D_1 = P_{10}$; $D_2 = P_{20}$; etc. $Q_2 = D_5 = P_{50}$ is a common relationship.

7. Semi-Interquartile Range

he *semi-interquartile range* (Q) *eliminates scores below* P_{25} *and above* P_{75}, and is another way of looking at range. The resulting number is *divided by two.* Using $P_{75}(Q_3)$ and $P_{25}(Q_1)$ from the example listed in **(4)**, computation of Q is shown below.

---

(5)

$$Q = \frac{Q_3 - Q_1}{2}$$

$$Q = \frac{66.21 - 54.89}{2}$$

$$Q = 5.66$$

Sometimes high and/or low scores will cause the data to appear to have a different measure of central tendency than it would have if an unusual score or two were disregarded. There are times when a gifted or low achieving person, for whatever reason, records a score very different from the rest of the students. By eliminating all data at either end of an otherwise normal distribution, a better understanding of what the numbers reveal may be obtained. Comparing Q in different distributions is one way of eliminating this kind of bias.

It is easy to confuse Q with quartile ($Q_?$). A quartile is usually listed with a subscript, such as Q_2 or described as the second quartile. On the other hand, Q is often written as semi-interquartile range or just Q.

8. Percentile Rank

This statistic is a *percent representing scores that are equal to or less than the given score*. It is found by the following procedure:

Where:

$$(6) \qquad PR_? = \frac{\left[\dfrac{(S-ll)}{i} * f_m\right] + CF_b}{N} * 100$$

S = score.
ll = lower limit of interval where score lies.
i = interval.
f_m = number of scores in the row where the score lies.
CF_b = cumulative frequency in the row directly below the row where the score lies.

Example: Using the data from **Table 7.3**, a percentile rank of a score of 54 is found by:

$$PR_{54} = \frac{\left[\dfrac{54-53.5}{5} * 9\right] + 11}{54} * 100 = 22.03\%$$

APPLICATIONS

If a score of 54 was made on a test, the PR of 22.03% is interpreted as 22.03% made a score of 54 or lower. Note the percentile rank is listed as a percent (%), not as a score. Recall that an ogive curve, as mentioned in Chapter 7, is another way to find a percentile rank or cumulative percent.

MEASURES OF CENTRAL TENDENCY

Measures of central tendency are reported in most data analyses. Look carefully to see if the distribution appears to be normal. Check the data classification and use the measure that is most appropriate. Consider normality, extreme scores and the following characteristics.

Mean

The mean is the best measure of average performance considering **all** score values with at least interval data. Use the mean with caution for comparisons of two similar groups on the same test. It must be remembered that extreme scores may *skew* this statistic adversely. There is an assumption of normality for the mean.

Median

**A** median is used to determine the middle point of performance, disregarding score values. Extreme scores will not affect the value of this statistic. Data should be at least nominal. Normality is not assumed.

Mode

The mode may be used to compare score homogenity in nominal data or above. It is common to use the mode to determine the position of most scores, rather than the value. No assumptions are made, other than the data are countable.

MEASURES OF POSITION

Percentile and percentile rank are used together to help establish position on a test. It is often customary to include ranges of scores or percents due to errors that may be asociated with any kind of measurement.

It is easy to confuse what is meant by percentile and percentile rank. Understanding the definitions and a careful study of the normal curve in Chapter 7 will help alleviate this problem.

Percentile

A percentile helps to identify a score which is equal to or less than a given percent. A person can learn their position if it is explained that 50% of the scores were 75 or less. National tests such as ACT or SAT may report scores and their corresponding percentile for that score. For example, an ACT English score of 25 may be shown with a percentage of 84, indicating that 84% of the test takers made a score of 25 or less on English. Human performance scores like Physical Best are reported and used in the same way.

Percentile Rank

The percentile rank shows the percentage (%) associated with a particular score. In the example above, a percentile rank of a score of 25 was 84%. If ranges of percentile are shown, such as 84% with ranges of 75% - 90%, it simply means that the test has an error range associated with the score so that the true percentage may vary from 75-90% on a similar test.

Decile, Quartile

At times scores may be designated into groups such as deciles or quartiles. For example, the fourth decile (D_4) indicates the percentile rank ranges from 40% through 49%. In a similar fashion, Q_2 represents a position from P_{26} to P_{50}.

These measures of central tendency and position are basic tools used in measuring groups of data. Although computer technology allows for quick solutions to these and other statistics, a knowledge about group solutions may help in understanding better these same numbers when all of these scores are used. Also, when displaying a lot of information on a single page, the grouped data equivalents of raw score data may be computed, rounded and used for approximations. Some major personal computer software include grouped as well as ungrouped solutions.

Chapter 9 will explore grouped data variability as well as standard scores.

EXERCISES

1. a.   Using **Table 8.1** below, make a frequency distribution table of the number of clients registering for new programs **Use i** = **5** and include CL, tally, f, Cum. f, Cum. Rel.. f, PR, d, fd and fd^2.

Table 8.1. Number of Clients Registering in New Programs Last Year.

161	194	173	164	166	164	178	181
178	167	184	145	196	184	185	162
190	150	159	190	183	207	192	162
184	156	169	165	157	160	205	185
147	190	184	186	176	197	165	173
183	156	165	192	187	176	186	178
179	179	204	207	174	184	214	193

   b.   Draw a frequency polygon and an ogive curve.
   c.   Estimate $PR_{184,\ 192}$ from the ogive.

2.   Using the data in exercise 1 above, find the following statistics:

   a. $\bar{X}$.   f. Semi-interquartile range.
   b. Md.   g. PR_{184}.
   c. Mode.   h. PR_{192}.
   d. P_{75}.   i. What is the AM?
   e. P_{25}.

3.   Assuming a normal distribution, are the following statements true (T) or false (F)? Explain your answers.

   a. $P_{50} = D_5 = Md$.
   b. Q_{25} and Q_{75} are equidistant from the median.
   c.   The median is exactly halfway between P_{20} and P_{80}.
   d.   The mode, median and mean are all the same score.

4.   A community TR program involved people with the following ages:

Table 8.2. Ages of Clients With Special Needs.

Ages	f
74 - 76	3
71 - 73	6
68 - 70	8
65 - 67	10
62 - 64	9
59 - 61	7
56 - 58	6
53 - 55	5
50 - 52	3
47 - 49	4
44 - 46	0
41 - 43	0
38 - 40	1

a.   Develop a frequency distribution table using columns CL, f, Rel. f, % of f, Cum. f, Cum. Rel. f, PR, d, fd and fd^2.
b.   Find $P_{10, 20, 50}$.
c.   Compute $PR_{53, 60, 71}$.
d.   What is the mode score?
e.   What is the median score?
f.   What is the mean score?
g.   Draw a bar graph and a line graph.
h. D_{10} includes scores with what P ranges?
i. Q_1 includes scores with what P ranges?
j.   In the row 38-40, what are the lower and upper limits?

5.   Given the following distribution:

Table 8.3. Fictitious Distribution.

CL	f	Cum. f
45 - 49	1	31
40 - 44	3	30
35 - 39	6	27
30 - 34	3	21
25 - 29	10	18
20 - 24	5	8
15 - 19	1	3
10 - 14	2	2

     a.   What is the assumed mean (AM)?
     b.   What is the mode score?
     c.   What is the median score?
     d.   What is the semi-interquartile range?
     e.   Find PR_{22}, P_{50}.
     f.   Draw a line graph.

6.   The following ages of participants using a fitness center in February were recorded:

Table 8.4. Ages of Clients Using Center in February.

CL	f	Cum. f
65 - 74	107	392
55 - 64	32	285
45 - 54	46	253
35 - 44	44	207
25 - 34	163	163

6. (continued)

   a. What is the assumed mean?
   b. What is the mode score?
   c. What is the median score?
   d. Find P_{50}.
   e. Find PR_{67}.
   f. What is the score for D_4?
   g. Find the score for Q_1.
   h. Find the semi-interquartile range.

7. Given the following data (U.S. Department of Commerce, 1992):

Table 8.5. International Revenue Passenger Miles, 1990, in Billions.

Month	Revenue Passenger Miles (Billions)
January	8.57
February	7.17
March	8.87
April	8.86
May	9.60
June	11.40
July	12.59
August	13.29
September	10.58
October	9.55
November	8.15
December	9.08

a.   Develop a frequency distribution table using columns CL, f, Rel. f, % of f, Cum. f, Cum. Rel. f, PR, d, fd and fd^2. **Use i = .51; top interval = 13.04-13.54.**

b.   Find $P_{10, 20, 50}$.

c.   Compute $PR_{7.17, 11.40, 13.29}$.

d.   What is the mode score?

e.   What is the median score?

f.   What is the mean score?

g.   Draw a bar graph and a pie graph.

h.   In the top row, what are the lower and upper limits?

8.   Given the following data (U.S. Dept. of Commerce, 1992):

Table 8.6. Revenue Passenger Miles, 1990.

Month	International Revenue Passenger Miles (Billions)	Total Industry Revenue Passenger Miles (Billions)
January	8.57	34.09
February	7.17	31.88
March	8.87	38.63
April	8.86	36.89
May	9.60	37.54
June	11.40	41.75
July	12.59	44.38
August	13.29	47.10
September	10.58	36.86
October	9.55	37.84
November	8.15	34.79
December	9.08	36.18

8. Using the Total Industry column, determine the following statistics:

a. Develop a frequency distribution table using columns CL, f, Rel. f, % of f, Cum. f, Cum. Rel. f, PR, d, fd and fd^2. **Use i = 1.5; top interval = 46.4 - 47.8.**

b. Find $P_{10, 20, 50}$.

c. Compute $PR_{34.09, 36.89, 37.84}$.

d. What is the mode score?

e. What is the median score?

f. What is the mean score?

g. Draw a mixed graph, using Total Industry data for the line.

Chapter 9

Variability and Standard Scores

VARIABILITY

1. Sum of Deviations squared

tandard deviation may be defined as the *square root of the mean of the sum of squares of the individual deviations from the mean.* It is denoted by the symbols s or sigma (σ). When used with other measures, it will show relationships of scores in one distribution to another, and is a major statistic reported in the literature.

Most discussions about standard deviation start with a statistic called the variance. Although the variance is an appropriate beginning, this text will begin with deviations, proceed to the variance and end with standard deviation.

A concept developed in the very beginning is that of the sum of deviations squared (Σx^2). It may be *defined as the square of individual score deviations from the mean.* This statistic is used in computing the variance, standard deviation, correlation coefficient, "t" testing and analysis of variance (ANOVA). The Σx^2 is commonly referred to as SS, but in most texts the emphasis is on other procedures. Since an understanding of Σx^2 is basic to so many other statistical tools, the beginning student should master the simple ways of computing this important number.

The procedure used in this text to determine sum of deviations squared, variance and standard deviation in grouped data is also applicable to raw scores. First, scores are placed in an array (column), with the high score at the top and low score at the bottom. In leisure services or tourism, time is often a measure used, and it must be recalled that the higher value is the slowest time (ie, 1 hour is a lower score than 2 hours, in most cases). Therefore, time is a special case, and scores usually need to be ranked with the smaller numbers (that represent higher values) at the top when analyzing time. Regardless of semantics, the score with higher value is always placed at the top of the column.

The *sum of deviations squared* is found by *subtracting the mean from each score, squaring the difference (x or deviation score) and summing the column*, as shown in **Table 9.1**:

Table 9.1. Sum of Deviations Squared.

fd	$\overline{X}$	$X-\overline{X}$ (x)	x^2
5	3	+2	4
4	3	+1	1
3	3	±0	0
2	3	-1	1
1	3	-2	4
			$\Sigma 10$

In the example above, Σx^2 is 10. But the same result can be obtained without having to compute the mean, subtract it from every score, square the result and sum the column. By using sums of the fd and fd^2 columns, the sum of deviations squared can be computed with fewer computations, *with less chance for error*. **Table 9.2** below shows the table needed, while equation 7 computes the same result.

Table 9.2. Σfd & Σfd^2.

fd	fd^2
5	25
4	16
3	9
2	4
1	1
$\Sigma 15$	55

$$\Sigma x^2 = \left[\Sigma fd^2 - \frac{(\Sigma fd)^2}{N}\right] * i^2$$

(7)
$$\Sigma x^2 = \left[55 - \frac{(15)^2}{N}\right] (1)^2$$

$$\Sigma x^2 = \left[55 - \frac{225}{5}\right] (1)$$

$$\Sigma x^2 = 55 - 45; \ \Sigma x^2 = 10$$

Note from equation 7 that the same result (10) is obtained by using the sums of two columns that was obtained in **Table 9.1**. With such a small data base, it may be easier to find the deviations and square them, as **Table 9.1** shows. However, when working with a large data base, say 25 or more scores, it is faster and usually more accurate to use the method described by equation 7. Furthermore, a modified form of equation 7 is a first step in analyzing ungrouped data (raw scores) to find the sum of deviations squared. As a further consideration, decimals may be minimized when using the equation method.

Since there are fewer numbers to manipulate, the equation 7 method will be used in this text to compute the sum of deviations squared. The rest of the procedure for finding variance and standard deviation is identical to the way followed when using raw scores only.

2. Variance

Variance (s^2) is defined as *an average deviation squared.* Therefore, all that is necessary to do is to divide the sum of deviations squared already obtained by N. In the case of a total population, or numbers from *all* subjects in a population, this statement is true. However, most statistical data does not include all of the possible scores.

When only a portion of the population scores are analyzed, the data is said to be a *sample* of that population. Therefore, an assumption made in this text is that the data is a sample, and instead of using N, N will be modified to N-1, which will be explained at a later time. To find s^2, divide the sum of deviations squared by N-1. An example with the data from equation 7 is shown below.

$$s^2 = \frac{\Sigma x^2}{N-1}$$

(8)

$$s^2 = \frac{10}{4}$$

$$s^2 = 2.5$$

3. Standard Deviation

The *standard deviation is the square root of the variance.* For a population, the symbol sigma (σ) is used, and for sample data, the symbol (s) is employed. Using the data from equation 8, the standard deviation is shown below.

(9)

$$s = \sqrt{s^2}$$

$$s = \sqrt{2.5}$$

$$s = 1.58$$

These three steps all assume sample data. If population scores were analyzed, the only change would be to substitute N for N-1. By using this method, either calculator, spreadsheet or data base information can be employed quickly and efficiently, with less chance for error. For example, when using a calculator, the steps can flow from start to finish without having to reenter previous data. A summary of the steps used to find s is shown in equation 10 below.

(10)

$$1.\ \Sigma x^2 = \left[\Sigma f d^2 - \frac{(\Sigma f d)^2}{N}\right] * i^2$$

$$2.\ s^2 = \frac{\Sigma x^2}{N-1}$$

$$3.\ s = \sqrt{s^2}$$

4. Standard Error of the Mean

The standard error of the mean is a vital concept. In sampling, the true mean (μ) is not known. Only when every score from a population has been gathered is the true mean known. Therefore, a sample mean ($\bar{x}$) has some error associated with it. The standard error of the mean (SEM or $s_{\bar{x}}$) identifies the error range of the sample mean.

The SEM is defined as s divided by the square root of N. Using the data from equation 9, the SEM is found in equation 11 to be:

$$s_{\bar{x}} = \frac{s}{\sqrt{N}}$$

$$s_{\bar{x}} = \frac{1.58}{\sqrt{5}}$$

(11)

$$s_{\bar{x}} = \frac{1.58}{2.24}$$

$$s_{\bar{x}} = .71$$

The error range may be interpreted as ± .71 applied to the mean. Since the sample mean was 3, the true mean may lie between 2.29 and 3.71 about 68% of the time. With only 5 scores to influence this mean, the error range appears to be fairly low.

5. Confidence Interval (CI) Testing

However, it is expedient to try to be more accurate. For reasons that deal with probabilities, it might be better to describe the error ranges within a specific probability. Most researchers use error ranges that have a 95% or 99% probability. At first glance, it would seem best to use the higher (99%) probability. Later it will be shown that as probability is increased, there will be a greater chance of making an alpha or Type I error. For the time being, the focus will be on interpreting the statistic without considering whether or not an alpha (α) or beta (Type II - β) error may be made.

To test at the .05 level (95% probability), multiply the standard error of the mean by 1.96 (two-tailed test). Add and subtract this product to and from the mean to determine error ranges at the .05 level. The 1.96 comes from the area under the normal curve that contains 95% of the scores (when using a two-tailed test, or both plus **and** minus sides of the normal curve). To test at the .01 level, multiply the mean by 2.58 (two-tailed test) for the same reasons. An example follows below in equation 12 using data from equation 11.

$$CI_{05} = \bar{X} \pm (1.96) * \frac{s}{\sqrt{N}}$$

(12)
$$CI_{05} = 3 \pm (1.96)(.71)$$

$$CI_{05} = 3 \pm (1.39)$$

$$CI_{05} = 1.61 - 4.39$$

Now the data appear slightly different. It seems the mean could be anywhere from 1.61 to 4.39 at the .05 level (or 95% of the time). With this sample mean of 3, the variation of where the true mean may lie is very high, and may be a good indication that the data are not very reliable. Without considering α or β errors, the .01 level (or 99% of the time) has an even higher error (1.16 - 4.83) associated with it.

One reason the error range is so high is because the number of scores sampled (N) is so small. It follows that to reduce the error range, all that is necessary is to increase N. If the standard deviation remains the same, and N is increased, the standard error of the mean will decrease. The reason for this occurrence is due to the procedure for determining the standard error (s÷$\sqrt{N}$). As N increases, the value of the denominator goes up too, so that the standard error decreases. It would be well to remember when gathering any sample data to make the sample size as large as possible in order to decrease the mean error range. In the above example, if $\bar{x}$ and s did not change, and N was increased from 5 to 31, the new SEM would be 1.58 (s)÷5.57 (square root of 31) or .28. The new CI_{05} becomes 3±(.28)*(1.96) or 2.45 - 3.55, a more acceptable 95% level.

Often measures in leisure services, sport management and travel are frequency counts (number of clients, number of programs, etc.) If these kinds of data are obtained, it may be expedient to have at least 31 subjects or scores for a test group, because these kinds of data are a chi-square distribution. In a chi-square distribution, the frequency curve approximates normality with 31 or more scores. Perhaps using

31 or more scores will give results that look like a normal curve. Draw a frequency curve of the data to see if it looks by inspection to approximate normality. There are other statistical methods, such as skewness formulas, to assist in determining the same thing, but simply drawing the frequency graph of the data may help in analysis.

STANDARD SCORES

T he concept of standard deviation may be used to compare data from different scales, based on the normal curve or "curve." Scores can be converted to standard "z" or "T" scores. Positions of these scores from different scales (time & frequency counts, for example) can then be compared.

1. z Scores

A z score converts a raw score to units of standard deviation for comparison purposes (see page 101). For example, time spent recruiting and number of clients recruited cannot be compared directly, because the scales of time and frequency count are different. Neither do they have the same means or standard deviations. However, when the data are converted to standard scores, they reflect relative positions of the scores in different distributions, so the scores of each recruiter can be added and averaged, and the result contains more information than either distribution considered by itself.

A *z score* is defined as a *deviation score (x) divided by the standard deviation.* z scores have a mean of 0 and a s of 1. To convert a raw score to a z score, follow the procedure listed below:

1. determine the mean score.
2. subtract $\bar{x}$ from **X** to obtain a deviation score (x).
3. divide by the standard deviation. Round to two decimal places.

In grouped data, the frequency distribution chart can be adapted for this purpose. Take the midpoint of the interval as the individual score and follow the procedure on the previous page. The computed z score now has a mean of zero and a standard deviation of 1. This result is why different measures using this system can be compared (z scores in all distributions have a mean of 0 and a standard deviation of 1).

Rather than having different means and standard deviations, scores from different data are converted to the same mean (0) and standard deviation (1) on each measure. An example of a fictitious distribution is shown below (when the interval is 1, real limits are from .5 to .9, as described previously, and the midpoint of an interval of 1 is the CL).

Table 9.3. Standard Scores of a Fictitious Distribution.

CL	X	$\bar{X}$	x $(X-\bar{X})$	s	z (x/s)	$10z$	T $(+50)$
7	7	4.14	2.86	2.64	1.08	10.82	61
6	6	4.14	1.86	2.64	0.70	7.03	57
5	5	4.14	0.86	2.64	0.32	3.25	53
4	4	4.14	-0.14	2.64	-0.05	-0.54	49
3	3	4.14	-1.14	2.64	-0.43	-4.33	46
2	2	4.14	-2.14	2.64	-0.81	-8.12	42
1	1	4.14	-3.14	2.64	-1.19	-11.90	38

2. T Scores

s shown in **Table 9.3,** a T score is an extension of a z score. To find a T score, follow the steps listed below:

1.   determine the z score.
2.   multiply z by 10.
3.   add 50 to the product. Round to the nearest whole number.

T scores have a mean of 50 and a standard deviation of 10. They are used the same way as z scores in comparing positions of numbers.

3. Using Computer Programs With Grouped Data

A word of caution is necessary when using computer technology to solve grouped data problems. If the program has been written to make the low score the midpoint of the lowest interval, the data will not coincide with data that uses the high score as the midpoint of the top interval. Since grouped data requires most of the analyses to start at the lowest interval and move upwards, it makes sense to use the low interval as a starting point to construct a frequency chart.

However, when starting at the lowest interval to construct the chart, it is necessary to count the number of rows needed and leave enough space to construct the entire chart. If the top interval is the starting point, it is very easy to work downward to complete the chart without first counting the rows needed. Since the data are grouped, some accuracy has been lost anyway. That is why this text uses the high score rather than the low score to organize data.

Grouped data statistics are not as easy to understand as raw score data, because the information has been organized into class intervals. For this reason, the data is less accurate than when using all of the scores without grouping. Although computer programs are available for grouped data, raw scores are more easily utilized in most computer work. Chapter 10 will analyze the statistics discussed so far by organizing the data without grouping (raw scores).

APPLICATIONS

1. Sum of Deviations Squared (Σx^2 or SS)

This important statistic is used as the first step in finding variance, standard deviation, correlation coefficient, "t" testing, and analysis of variance. It is found by subtracting the mean from each score, called a deviation or (x) score and squaring the deviations (x^2). All of the deviations are then summed to produce the sum of deviations squared (Σx^2). When using the sum of scores (ΣX) and sum of scores squared (ΣX^2) columns to find the sum of deviations squared, the student is encouraged to visualize what the x and x^2 scores represent.

2. Standard Deviation (s)

Many of the more advanced statistical techniques are based on measures of variability, of which standard deviation is a major tool. Standard deviation is the best measure of variability. It is very useful for measuring variation of individual scores in two comparable groups on the same data scale (with reservations when extreme scores are not influential). Many research articles, such as those found in the *Journal of Leisure Research* or *TTRA Journal* contain data that includes standard deviation. A special tool not explained in this text called meta analysis compares different studies with different numbers of subjects on standard deviation.

3. Standard Error of the Mean (SEM)

The standard error of the mean is used to determine the error range of data that a sample mean has. Depending on the data, a large error range may be interpreted as not representative of the true mean (μ). When the SEM is not multiplied by a z score representing a percentage, it indicates the error range 68% of the time.

The SEM can be lowered by increasing N in the sample size. Recall that in frequency counts, larger frequencies begin to look more like a normal curve. A random sample of 31 or more scores may resemble a normal curve. Therefore, the closer N is to 31, the more likely the data is to resemble a normal curve.

4. Confidence Interval (CI)

The CI is used with the SEM to determine error range of a sample mean at desired confidence levels. By multiplying the SEM by a z of 1.96, a confidence interval of 95% (.05 level, 2 tailed test) is obtained. In the same way, using a z of 2.58 results in a 99% (.01) confidence level (2 tailed test).

The .01 level may not be any more desirable than the .05

level, because as confidence levels increase, chances for making a Type II or Beta (β) error increase in what is called confidence interval testing. Confidence intervals show the error range of the sample mean at desired percentages.

5. Standard Scores (z, T)

Standard scores are used in evaluating data when the position of one score from one distribution is to be compared with other positions from the same or different distributions. Since standard scores convert the data to the same mean and standard deviation, different kinds of tests, such as performance tests with written tests, can be compared more efficiently. However, there is nothing absolute about a standard score. It should be used with other data for final interpretation.

Grouped data are not used frequently today. However, reports that use many pages of data may be more easily understood if the data are grouped and reported on one page (frequenct distribution chart) where analysis is easier to recognize. By grouping the data and rounding numbers to a more easily understood framework, gaps in data can be discovered more easily

Part 3 introduces statistical concepts where all the scores are used. Chapter 10 reviews measures of central tendency, position and variability and explains how these tools are derivied when all the scores are used directly.

Exercises

The following table will be used for several exercises:

Table 9.4. Selected Data.

Program Number	Attendance Quarter 1	Attendance Quarter 2	Attendance Year
1	52	55	196
2	54	40	165
3	64	56	193
4	58	47	197
5	53	51	163
6	46	48	190
7	53	46	176
8	63	47	187
9	52	45	149
10	44	44	177
11	51	59	174
12	45	32	214
13	60	40	112
14	35	54	177
15	64	47	185
16	53	49	170
17	54	41	141
18	50	55	183
19	60	55	204

1.   a.   Using **Table 9.4**, Quarter 1, create a frequency chart **with i** = 3 including CL, f, Cum. f, PR, d. fd. fd^2, X, mean, x, s, z, 10z, T.

      b.   Determine the mean, mode and median.

      c.   Draw a frequency polygon and an ogive curve. Do the data appear normal? Why or why not.

      d.   Find P_{75}, P_{25}, Q.

      e.   What is PR_{35}, PR_{64}?

      f.   What is the Σx^2, s^2, $s_{\bar{x}}$, $CI_{.05}$, $CI_{.01}$?

2.   a.   Using **Table 9.4**, Quarter 2, generate a table **with i = 2** including CL, f, cum. f, PR, d, fd. fd^2, X, mean, x, s, z, 10z, T. For simplicity, **make the high score (59) the top of the interval (top CL = 58-59, real limits = 58.0-59.99; midpoint = 58.5)** and continue down.

      b.   Determine the mean, mode and median.

      c.   Draw a frequency polygon and an ogive curve. Do the data appear normal? Why or why not.

      d.   Find P_{75}, P_{25}, Q.

      e.   What is PR_{40}, PR_{55}?

      f.   What is the Σx^2, s^2, $s_{\bar{x}}$, $CI_{.05}$, $CI_{.01}$?

3.   a.   Using the Year Attendance scores in **Table 9.4**, create a table **with i = 7** including CL, f, cum. f, PR, d, fd. fd^2, X, mean, x, s, z, 10z, T.

      b.   What is the variance? $s_{\bar{x}}$? CI at .05 level?

4. The mean and standard deviation for free passes redeemed are 18.5 and 2.1 respectively. With an N of 31, find z and T for the following scores: 13, 17, 20, 23.

5. Complete the following table.

Table 9.5. Attendance at Metroplis Center.

CL	X	$\bar{X}$	$X-\bar{X}$	s	z	10z	T
89-93		60.44		11.19			
84-88		60.44		11.19			
79-83		60.44		11.19			
74-78		60.44		11.19			
69-73		60.44		11.19			
64-68		60.44		11.19			
59-63		60.44		11.19			
54-58		60.44		11.19			
49-53		60.44		11.19			
44-48		60.44		11.19			
39-43		60.44		11.19			
34-38		60.44		11.19			
29-33		60.44		11.19			

Part 3

Raw Scores

Central Tendency & Variability

Chapter 10

RAW SCORES

When using raw scores, there is no need to group data into class intervals. Instead, the numbers (scores) are usually ranked in descending order; that is, high score at the top and other scores underneath in descending order. A ranking of this nature is called an array. If a computer program is used, it may not be necessary to rank the data.

MEASURES OF CENTRAL TENDENCY

1. The Mean ($\bar{X}$)

The mean is the average score. It takes into account all scores in a distribution. Therefore, extreme scores may affect this statistic.

To determine the mean, rank all the data from high to low, add the total and divide by N, as in **Table 10.1** and Equation 13.

Table 10.1. Number of Baskets in 15 Seconds.

Rank	X
4.0	6
2.5	4
2.5	4
1.0	2
	$\Sigma\ 16$

(13)

$$\bar{X} = \frac{\Sigma X}{N}$$

$$\bar{X} = \frac{16}{4}$$

$$\bar{X} = 4$$

In the example above, the mean is 4. However, one extreme score can score can affect the mean considerably. Using the same distribution, but adding one extreme score as **in Table 10.2** results in the following data:

Table 10.2. One Extreme Score (18).

Rank	X
4.0	18
2.5	4
2.5	4
1.0	2
	Σ 28

(14)

$$\bar{X} = \frac{\Sigma X}{N}$$

$$\bar{X} = \frac{28}{4}$$

$$\bar{X} = 7$$

Now the mean is considerably different from the mean in Equation 13 even though only one score has changed. It is important to inspect all of the scores for extremes at either or both ends. Characteristics of the mean were discussed in Chapter 8.

2. Median (Md)

he median is the middle score in a distribution. Fifty percent of the scores lie above and below the median. The median *rank* is found by using (N+1)÷2. Since both examples in **Tables 10.1** and **10.2** have an N of 4, the median rank for both distributions is 5÷2 = 2.5. **Table 10.3** shows that starting at the bottom of either array and going up to find a rank of 2.5 results in a median score of 4 in each case.

Table 10.3. The Median in Two Distributions.

Distribution A		Distribution B	
Rank	X	Rank	X
4	6	4	18
3	4	3	4
→(2.5) 2	→(4) 4	→(2.5) 2	→(4) 4
1	2	1	2

The ranks of 2.5 in each distribution in **Table 10.3** are found by starting at the lowest rank and proceeding up the rank column until the desired rank (2.5) is found. If a rank falls between two listed ranks, go halfway between the ranks to locate the desired rank (in the example given in **Table 10.3**, the rank of 2.5 has been added in parentheses). The median score (X) is the score that is on the same row as the desired rank of a score of 4 (in the example, the score of 4 has been added in parentheses). Even though distribution B has an extreme score of 18, the median is unaffected. That is why the median may be a better measure of central tendency with extreme scores.

When using raw scores (all the scores), it is easy to find the median.

1.　Place all scores in a column, high to low, top to bottom.

2.　Find the median rank by $(N+1) \div 2$.

3.　Locate the *score* that corresponds to this rank. It is in the same row as the median rank. If the *score* lies between two scores that do not have the same value, add the adjoining scores an divide by 2 to arrive at the median score, as shown in **Table 10.4**.

Table 10.4.Median Between Two Scores.

Rank	X
6	9
5	8
4	6
(3.5)	> ?
3	4
2	2
1	1

Since N=6, the median rank is $(6+1) \div 2 = 3.5$ (the rank of 3.5 has been added in parentheses for emphasis).The median rank in **Table 10.4** falls halfway between a score of 4 and a score of 6, as shown by the question mark in the X column. Adding the two adjacent scores, 4+6=10 and dividing by 2=5. Therefore, a median rank of 3.5 in **Table 10.4** results in a median *score* of 5. Characteristics of the median were discussed in Chapter 8.

3. Mode (Mo)

T he mode score when using raw data is the same as when using grouped data (see Chapter 8). It is the *score* of highest frequency, or the number that occurs the most times. If more than one mode is found, list them all. If there are not two or more scores with the same value, there is no mode.

VARIABILITY

The variance (s^2) is a statistic that is used in many kinds of data analyses. It is a part of standard deviation, standard scores, "t" tests, ANOVA, and many other techniques, including non-parametric statistics. Variance is based upon distance from the mean. It is the sum of the squared deviations from the mean divided by N-1 (or N in the case of a total population distribution).

1. Sum of Deviations Squared (Σx^2)

 s with grouped data, the process begins with the sum of deviations squared. Since the data are not grouped, the column equivalents of fd and fd^2 are used, and since the scores were not coded, there is no interval. The column equivalent of fd is X, and the column equivalent of fd^2 is X^2.

The procedure for finding Σx^2 is:

(15)
$$\Sigma x^2 = \Sigma X^2 - \frac{(\Sigma X)^2}{N}$$

An example of a sample distribution in **Table 10.5** (next page) may help to serve as a reminder as to how the sum of deviations squared is found *without* using equation 15. This procedure was also used in Chapter 9 with grouped data.

Table 10.5. Finding Sum of Deviations Squared.

X	$- \bar{x}$	$= x$	x^2
5	3	+2	4
4	3	+1	1
3	3	±0	0
2	3	-1	1
1	3	-2	4
			Σ 10

In **Table 10.5** the sum of deviations squared is 10. The method used to find Σx^2 may be termed *definitional,* since the procedures follow the definition of what is meant by the sum of deviations squared. Notice how this statistic was obtained. A mean of all the X scores is found, and this mean is subtracted from each score in the X column. The result is a *deviation* (x) from the mean. The deviations are then squared, and the sum of the squared deviations is computed. As simple as this process is, the Σx^2 is a key to understanding other statistical tools. It is usually symbolized by SS in analysis of variance.

With a few scores, the procedure shown in **Table 10.5** is fast way to compute Σx^2. When using certain computer software that can quickly copy formulas into cells, the method may be satisfactory. But it may be a lot easier to find Σx^2 in a different way that will eliminate finding deviations (x).

If the definitional method is used, the deviation determined will often be a decimal. A long column of decimals is tedious to compute. Even software that allows column formats takes some time, and there may be a slight loss of accuracy in the process. Then, the x scores need to be squared, with a need to round decimals once again, with perhaps an additional slight loss in accuracy. Whereas software can quickly sum even a long column of decimals accurately, other methods fall short of the precision that can be obtained in the manner described.

The Σx^2 can be found by using the procedure defined in equation 15. In order to use this approach, all that is needed is two columns, the X (raw) scores and these scores squared (X^2). **Table 10.6** below uses the data from **Table 10.5** as the basis for finding Σx^2.

Table 10.6. Sum of Deviations Squared.

X	X^2
5	25
4	16
3	9
2	4
1	1
Σ 15	55

(16)

$$\Sigma x^2 = \Sigma X^2 - \frac{(\Sigma X)^2}{N}$$

$$\Sigma x^2 = 55 - \frac{(15)^2}{5}$$

$$\Sigma x^2 = 55 - \frac{225}{15}$$

$$\Sigma x^2 = 55 - 45 = 10$$

Notice the data in equation 16 bypasses x scores and computes Σx^2 directly. Using this method eliminates the intermediate step of finding x scores, which usually are decimals, and requires creating only 1 additional column (X^2), whereas using the definitional procedure in **Table 10.5** requires at least 2 additional columns (x, x^2). This text will use the method shown in equation 16 above for all computations involving sum of deviations squared. However, careful study of **Table 10.5** is suggested in order to understand what the sum of deviations squared really means.

2. Variance (s^2)

The variance using raw scores is found the very same way as the method used with grouped data already discussed in Chapter 9. The variance is a major statistic, and is an intermediary step in computing the standard deviation. Recall that this text assumes the data are sample numbers and not population figures. Sample data uses N-1 under the radical, while population data uses N for the denominator. It is important to use the proper designator of N or N-1 in order to more accurately find appropriate significance levels.

Using data in **Table 10.6** and equation 16, the sample variance is:

(17)

$$s^2 = \frac{\Sigma x^2}{N-1}$$

$$s^2 = \frac{10}{5-1}$$

$$s^2 = \frac{10}{4} = 2.5$$

3. Standard Deviation (s)

he standard deviation using raw scores is found the same way as in grouped data. Using equation 17, the standard deviation is:

(18)

$$s = \sqrt{s^2}$$

$$s = \sqrt{2.5}$$

$$s = 1.58$$

Comparing Grouped Data With Raw Scores

t may be useful to compare relationships between the procedures used in grouped and ungrouped data. **Table 10.7** on the following page is a distribution of scores from a written pre-test.

Table 10.7. Pre Test Scores.

38	64	50	32	44	25	49	57	46	58
40	47	36	48	52	44	68	26	38	76
63	19	54	65	46	73	42	47	35	53
40	35	61	45	35	42	50	56	45	28

The scores are arranged in Figure 10.1 to show both sets of data.

Grouped Data					Raw Scores		
CL	f	d	fd	fd²	Rank	X	X²
74 - 78	1	6	6	36	40	76	5776
69 - 73	1	5	5	25	39	73	5329
64 - 68	3	4	12	48	38	68	4624
59 - 63	2	3	6	18	37	65	4225
54 - 58	4	2	8	16	36	64	4096
49 - 53	5	1	5	5	35	63	3969
44 - 48	9	0	0	0	34	61	3721
39 - 43	4	-1	-4	4	33	58	3364
34 - 38	6	-2	-12	24	32	57	3249
29 - 33	1	-3	-3	9	31	56	3136
24 - 28	3	-4	-12	48	30	54	2916
19 - 23	1	-5	-5	25	29	53	2809
					28	52	2704
	Σ 40			6 258	26.5	50	2500
					26.5	50	2500
					25	49	2401
					24	48	2304
					22.5	47	2209
					22.5	47	2209
					20.5	46	2116
					20.5	46	2116
					18.5	45	2025
					18.5	45	2025
					16.5	44	1936
					16.5	44	1936
					14.5	42	1764
					14.5	42	1764
					12.5	40	1600
					12.5	40	1600
					10.5	38	1444
					10.5	38	1444
					9	36	1296
					7	35	1225
					7	35	1225
					7	35	1225
					5	32	1024
					4	28	784
					3	26	676
					2	25	625
					1	19	361
					Σ	1872	94252

Figure 10.1. Comparison of Grouped and Raw Scores.

Data from **Figure 10.1** is used to compare Σx^2, s^2, and s in the equations below.

Grouped Data	*Raw Scores*

1. $\Sigma x^2 = \left[\Sigma fd^2 - \dfrac{(\Sigma fd)^2}{N}\right](i)^2$ **1.** $\Sigma x^2 = \Sigma X^2 - \dfrac{(\Sigma X)^2}{N}$

$\Sigma x^2 = \left[258 - \dfrac{(6)^2}{40}\right](5)^2$ $\Sigma x^2 = 94{,}252 - \dfrac{(1{,}872)^2}{40}$

$\Sigma x^2 = \left[258 - \dfrac{36}{40}\right](25)$ $\Sigma x^2 = 94{,}252 - \dfrac{3{,}504{,}384}{40}$

$\Sigma x^2 = 6{,}427.50$ $\Sigma x^2 = 6{,}642.40$

2. $s^2 = \dfrac{\Sigma x^2}{N-1}$ **2.** $s^2 = \dfrac{\Sigma x^2}{N-1}$

$s^2 = \dfrac{6{,}427.50}{39}$ $s^2 = \dfrac{6{,}642.20}{39}$

$s^2 = 164.81$ $s^2 = 170.31$

3. $s = \sqrt{s^2}$ **3.** $s = \sqrt{s^2}$

$s = \sqrt{164.81} = 12.84$ $s = \sqrt{170.31} = 13.05$

Several things should be noted about this comparison, including the similarity of the formulas used to obtain Σx^2, the variance (s^2), and the standard deviation (s). First, a discussion given in Chapter 7 revealed how the f, d and fd columns were obtained, and explained how the grouped data columns and procedures were used in much the same way as raw scores. The steps used above show how close the data analysis is in both grouped and raw scores.

By now it should be apparent that the fd and fd^2 columns in grouped data approximate the X and X^2 columns in raw scores. However, in grouped data the interval was divided into each midpoint to produce each d entry, and the d entry was squared to obtain the fd^2 column. Therefore, the interval needs to be squared and multiplied by the final result to obtain the Σx^2 for grouped data.

The other steps in finding the variance and standard deviation are the same for both grouped and ungrouped data.

Another discovery is that the standard deviations are not the same when comparing grouped and ungrouped data. This difference is due to the fact that the midpoints of the intervals were assumed to be the scores, whereas actual scores may or may not have been near the midpoint of grouped data. Also, data in the interval may not be homoscedastic.

Raw score data is certainly more accurate, but in this case, which is representative of general results, the difference is only .2 of a standard deviation (when rounded to the nearest tenth). If a larger N were used, the error could have been even less.

For a quick approximation of results, grouped data may be satisfactory, but with computer technology, it is usually more expedient and accurate to use raw scores for computations. Grouping data is still useful for displaying a large amount of data on a single page and for learning how computer graphics produce those nice looking curves and bar graphs.

Another thing to notice about the raw data is shown in the rank column on page 152. The ranks are needed when determining a percentile or percentile rank. In several cases, the rank listed is for a tied rank; that is, there are several scores that are the same value. For example, there are three scores of 35, and a rank of 7 is listed for them. The rank of 7 is obtained by adding the usual positions of the tied scores and dividing by the number of tied ranks.

Positions 6, 7 & 8 should have occupied the ranks for these scores. Since the scores are the same, they must all have the same rank. By adding the ranks together and dividing by the number of tied scores, the rank for the tied scores can be found. In this case, 6+7+8 = 21. Since there are 3 scores, 21÷3 = 7. Therefore, each of the scores has the rank of 7. The same procedure is used to find ranks of all tied scores.

For example, if only two ranks are tied, the rank sums are divided by 2; if four ranks are tied, the sum is divided by 4, etc. Notice that where the tied ranks end, the next rank up is *what it should be if no ranks were tied*; ie., a score of 36 is a rank of 9, which is the position it occupies right after a rank of 8.

Problem Solving

I t may be helpful to develop a table similar to a frequency chart when using raw scores. A table can help to organize the data so that standard deviation and standard scores can be quickly determined. Consider the following problem presented by attendance from 12 workshops:

$$63, 56, 36, 47, 84, 78, 69, 36, 56, 74, 39, 15.$$

First, rearrange the scores from high to low, just like a frequency chart. Then work out other statistics, such as the mean, mode, median, P_{75}, P_{25}, Q, Σx^2, s^2, s, z and T. Figure 10.2 summarizes a raw score chart from the data shown above.

Rank	X	Mean	X - Mean	s	z	10z	T	X²
12	84	54.42	29.58	20.52	1.44	14.42	64	7,056
11	78	54.42	23.58	20.52	1.15	11.49	61	6,084
10	74	54.42	19.58	20.52	0.95	9.54	60	5,476
9	69	54.42	14.58	20.52	0.71	7.11	57	4,761
8	63	54.42	8.58	20.52	0.42	4.18	54	3,969
6.5	56	54.42	1.58	20.52	0.08	0.77	51	3,136
6.5	56	54.42	1.58	20.52	0.08	0.77	51	3,136
5	47	54.42	-7.42	20.52	-0.36	-3.62	46	2,209
4	39	54.42	-15.42	20.52	-0.75	-7.51	42	1,521
2.5	36	54.42	-18.42	20.52	-0.90	-8.98	41	1,296
2.5	36	54.42	-18.42	20.52	-0.90	-8.98	41	1,296
1	15	54.42	-39.42	20.52	-1.92	-19.21	31	225
Σ	653							40,165

Figure 10.2. Raw Score Chart.

ANALYZING RAW SCORE DATA

Measures of Central Tendency

After the scores are arranged in an array from high to low, determine the rank of each score. Be sure to account for ties, with the remaining untied ranks in their proper order. Next, add up the X (653) and X^2 (40,165) columns. Note that N = 12.

1. The Mean

 he mean is found by adding all the scores and dividing by N. In the problem on the previous page, it is $653 \div 12$ or 54.42.

2. The Mode

 mode is the score of greatest frequency. There are two modes. One is at the rank of 2.5 (score of 36), and the other one is at rank 6.5 (score of 56). Therefore, the modes are 36 and 56.

3. The Median

 edian scores have exactly 50% of the total scores (N) both above and below them. The median rank is found by $(N+1) \div 2$. In the example cited, the median rank is $(12+1) \div 2 = 13 \div 2 = 6.5$. Fortunately, there is already a rank of 6.5 in the data, as shown by the score of 56. The median is the score in the same row opposite the median rank. Therefore, the median is 56.

If the median rank lies between two known ranks, an interpolation of the ratio between ranks and scores is needed. The procedure is the same as given later for P_{25}, so an example will be delayed until then.

MEASURES OF VARIABILITY

1. Sum of Deviations Squared

sing the data in **Figure 10.2** the sum of deviations squared is computed as:

$$\Sigma x^2 = \Sigma X^2 - \frac{(\Sigma X)^2}{N}$$

$$\Sigma x^2 = 40{,}165 - \frac{653^2}{12}$$

$$\Sigma x^2 = 40{,}165 - 35{,}534.08$$

$$\Sigma x^2 = 4{,}630.92$$

2. Variance

he variance is found by:

$$s^2 = \frac{\Sigma x^2}{N-1}$$

$$s^2 = \frac{4{,}630.92}{12-1}$$

$$s^2 = \frac{4{,}630.92}{11}$$

$$s^2 = 420.99$$

3. Standard Deviation

The standard deviation is determined by:

$$s = \sqrt{s^2}$$

$$s = \sqrt{420.99}$$

$$s = 20.52$$

MEASURES OF POSITION

1. Percentile, Decile, Quartile, Q

Percentiles, deciles and quartiles are found in much the same way as in grouped data (see Chapter 8). First, find what rank the percentile has. To find P_{25} in **Figure 10.2**, 12 X .25 = 3. Look at the ranks, find the rank of 3, and P_{25} is the *score* corresponding to the rank of 3.

Unfortunately, there is no rank of 3. There is a rank of 2.5, and a rank of 4. It is obvious a rank of 3 must fall somewhere between these two ranks, and a ratio will be needed to find the exact score of a position between two ranks. The same thing can be done if the median falls between two ranks. The following proportion may be used:

$$\frac{\textit{Difference Between Ranks}}{\textit{Difference Between Rank Wanted}} = \frac{\textit{Difference Between Scores}}{\textit{Score Difference}(X)}$$

Where Diff. Between Ranks = Above AND Below Needed Rank

Diff. Between Scores = Above AND Below Needed Score

Diff. Between Rank Wanted = Rank Wanted Minus Lower Rank

Score Difference = Value To Add To Lower Score

(19)

In the example given:

1. $\dfrac{4-2.5}{3-2.5} = \dfrac{39-36}{X}$

2. $\dfrac{1.5}{.5} = \dfrac{3}{X}$

3. $1.5X = 1.5$

4. $X = 1$

The result of equation 19 is a score difference that must be added to the lower score in order to find P_{25}. Since the lower score (just below the position where P_{25} lies) is 36,

$$P_{25} = 36 + 1 = 37$$

In the same way, P_{75} in Figure 10.2 is found:

$12 \times .75 = 9$. Since a rank of 9 is listed, $P_{75} = 69$.

To find Q:

(20)

1. $Q = \dfrac{P_{75} - P_{25}}{2}$

2. $Q = \dfrac{69 - 37}{2}$

3. $Q = \dfrac{32}{2}$

4. $Q = 16.0$

Since *deciles* D_1, D_2, etc. are *percentiles* (P_{10}, P_{20}, etc.), no examples will be given. They are found in the same way percentiles are determined. *Quartiles* are similar ($Q_1 = P_{25}$; $Q_2 = P_{50}$, etc.), and no examples of quartiles will be given either (see Chapter 8).

2. Percentile Rank

Using **Figure 10.2**, finding percentile rank for raw scores is easy when compared to grouped data. All that is necessary is to locate the score, determine the rank of the score, divide this rank by N, and multiply the result by 100. In the present problem, to determine PR_{74}, see equation 21 below. Interpolation, if needed, is the same as percentile in equation 19.

(21)

$$PR_{74} = \left[\frac{10 \ (Rank \ of \ PR_{74})}{12 \ (N)}\right] *100$$

$$PR_{74} = .833 * 100$$

$$PR_{74} = 83.3\%$$

3. Standard Scores

Standard scores are found in the same manner as they were found using grouped data:

1. subtract $\bar{x}$ from **X**.
2. divide by **s** for **z**.
3. multiply by 10.
4. add 50 for **T**.

Elementary statistics are important in understanding more complicated ways of analyzing data. The student would do well to review all basic concepts and complete all problems in the exercises given to this point. Chapter 11 will describe some elementary concepts about hypothesis testing that will be needed in analyzing data further.

APPLICATIONS

Since applications were given for measures of central tendency in Chapter 8 and for variability and measures of position in Chapter 9, no applications will be presented in this chapter. Please refer to the prior discussions in the aforementioned chapters.

EXERCISES

Reproduced below are **Tables 8.1 and 7.10.**

Table 8.1. Number of Clients Registered in New Programs Last Year.

161	194	173	164	166	164	178	181
178	167	184	145	196	184	185	162
190	180	159	190	183	207	192	162
184	156	169	165	157	160	205	185
147	190	184	186	176	197	165	173
183	156	165	192	187	176	186	178
179	179	204	207	174	184	214	193

1.　　　Use the data in **Table 8.1** above to find the statistics below:

a. $\bar{X}$.　　　　h. P_{75}.
b. **md.**　　　　i. P_{25}.
c. **mode.**　　　j. **Semi-interquartile range.**
d. $s_{\bar{X}}$.　　　k. PR_{184}.
e. Σx^2.　　　l. PR_{192}.
f. s^2.　　　　m. **What is the assumed mean?**
g. **s.**　　　　n. $CI_{.05, .01}$ **ranges.**

Table 7.10. Playground Attendance at 36 Sites.

29	50	36	28	54	49
33	53	40	38	37	56
35	57	47	46	41	51
43	29	55	30	52	43
44	25	26	32	58	39
27	31	42	48	54	45

2. Find the following statistics using **Table 7.10** on the previous
    page.

    a. $\bar{x}$.   i. P_{25}.
    b. **Md.**   j. **Semi-interquartile range.**
    c. **Mode.**   k. PR_{25}.
    d. $s_{\bar{x}}$.   l. PR_{50}.
    e. Σx^2.   m. **What is the assumed mean?**
    f. s^2.   n. $CI_{.05, .01}$ **ranges.**
    g. **s.**   o. **z, T scores.**
    h. P_{75}.

Table 10.8. Fictitious Data.

10.50	13.00	9.75	7.50	10.00	8.50	11.50
8.25	8.50	9.50	8.75	9.50	11.25	9.75
12.50	9.90	9.00	8.25	8.75	7.75	7.00
10.25	9.00	8.75	8.75	6.75	7.00	9.25

3. Using **Table 10.8**, determine the following statistics:

    a. $\bar{x}$.   i. P_{25}.
    b. **Md.**   j. **Semi-interquartile range.**
    c. **Mode.**   k. PR_9.
    d. $s_{\bar{x}}$.   l. PR_{13}.
    e. Σx^2.   m. **What is the assumed mean?**
    f. s^2.   n. $CI_{.05, .01}$.
    g. **s.**   o. **z, T scores.**
    h. P_{75}.

4.   Listed below are ages of clients on a tour.

Table 10.9. Ages of Clients on Tour.

58	74	67	65	68	57	64	55	48
52	65	60	50	59	71	66	60	39
61	62	71	68	57	56	63	61	52
50	65	58	62	72	56	70	72	47
40	51	70	73	73	68	70	66	67
72	48	64	53	69	61	63	49	76
58	63	75	66	62	59	59	67	

a.   What is the mean score?
b.   What is the median score?
c.   What is the mode score?
d.   Find PR_{49}.
e.   Find PR_{60}.
f.   What is $s_{\bar{x}}$?
g.   What are $P_{75,\ 25}$?
h.   What is the semi-interquartile range?
i.   Find a score of D_2.

5.   With a $s_{\bar{x}}$ of 8.3, what could be done to decrease this high error range on the next sample?

6.   Use a computer program to produce a bar graph for exercises 1 & 2 and a frequency polygon for exercises 3 and 4.

Chapter 11

Sampling and Hypothesis Testing

Very seldom is it possible to get all the data from an entire population. For example, if data from all clients in a spa are desired, it would be possible to go to a list of all client members in order to identify everyone. But the membership list is not likely to be complete. Surely a client has moved recently, and their name remains on the list. Or a new person has joined, is attending regularly, and this name is not on the list. The best of lists will probably not be completely accurate, and to whatever extent the list is inaccurate, the population statistics will not be appropriate.

If an entire population (σ) can be obtained, then *N rather than N-1* is used to determine variance. That is:

$$\Sigma x^2 = \Sigma X^2 - \frac{(\Sigma X)^2}{N}$$

(22) *Then:*

$$s^2 = \frac{\Sigma x^2}{N}$$

Also, μ = population mean, (whereas $\bar{x}$ = sample mean), and σ = population standard deviation.

Most statisticians use sampling procedures to gather data. In sampling, it is not necessary to use all the scores. Instead, a relatively small number of scores are obtained with the assumption that these scores represent population data.

Of course, there is some error in a sample, and the data are not the same as if the entire population set of scores were used. But it is assumed that the error is not large, and that is why looking at the mean and confidence interval helps determine the ranges of errors a statistician is willing to accept a certain percentage of the time.

1. Hypothesis testing

I n Chapter 9 it was stated that most statisticians use the .05 (95%) and .01 (99%) confidence intervals. These are not the only intervals that could be used, but they are recognized by most people as being the most convenient. Other levels, such as .10 (90%) could be set by the researcher analyzing the data.

These points (.05, .01) are found by examination of the normal curve. Raw scores are converted to standard or z scores that have a mean of 0 and an s of 1. The procedure for finding z is:

$$(23) \qquad z = \frac{X - \bar{X}}{s}$$

The z score represents a point on the line underneath the normal curve. Reference is made to **Figure 7.7** on page 101 in Chapter 7. Notice the z scores on the normal curve coincide with the standard deviation numbers directly underneath the curve.

The curve can be thought of as representing percentages of area. The entire curve represents 100% of the data (or area) under the curve. One standard deviation to the right of the mean represents 34.13% of the total area under the curve, 2 standard deviations represent 34.13% + 13.59% or 47.72% to the right of the mean, etc.

When the .05 level is used, 1.96 standard deviations (or a z of 1.96) represent 47.5% of the area in one direction. In hypothesis testing, the direction being tested may be known. For example, if a known fitness prescription generally produces an increase in fitness, and that method is used with subjects in gathering data, the expectation is an increase in fitness (or z scores) that would be on the plus side of a normal curve. In a situation where direction is known, the test could be on one side of the curve, in the direction expected. This test is known as a *one-tailed test*.

In the case of a one-tailed test at the .05 level, a z score or standard deviation unit of 1.64 represents 95% of the area from a z of 0 to + or - 1.64. Hence, to make a one tailed test at the .05 level,

1.64 units should be used as the constant, rather than 1.96.

When direction cannot be predicted, or when a more rigorous test of the data is desired, both the plus and minus sides of the mean are used. Therefore, the standard deviation unit or z score that contains 47.5% on each side of the curve needs to be found. This standard deviation number is 1.96, and is used to test the area on both sides of the curve (47.5 + 47.5 = 95% or the .05 level).

What confidence interval testing is doing is finding the standard error associated with a sample mean, defined as $s \div \sqrt{N}$, then multiplying that error times the confidence interval desired (ie, 1.96 = .05, 2.58 = .01). When this result is applied to the sample mean in both directions, the ranges of the mean are predicted. That is, either 95% or 99% of the time, the mean should be within the ranges found.

It should be clear from this reasoning that the sample mean does not have to fall within these ranges. All the confidence interval can do is predict that 95% or 99% of the time, the sample mean should be within the limits found. However, there is no way to know if the sample mean obtained is within the range, or whether it is one of the 5 or 1 in 100 times the sample mean is outside the predicted range.

2. Probability

Confidence interval testing leads to a discussion of probability. Rather than use mathematical concepts of N factorial, this text will confine the logic to a very simple idea; that is, it is more likely the sample mean is in the 95 of 100 times predicted than in the 5 of 100 times not predicted. There are more opportunities to be in the 95 of 100 group than in the smaller 5 of 100 group. Also, the area under the normal curve is referred to as a %, with all the area equal to 100%.

Tables that use the .05 and .01 levels are using this concept. Some tables list the values for a one-tailed test. When this method is used, halving the probability listed may provide the data for a two-tailed test. If the table lists probability for a one-tailed test, and a two-tailed test at the .05 level is needed, use the value listed for .025 rather than .05. For a .01 level two-tailed test, use the value for .005 instead.

It must be remembered that in sampling, there is no way to know if the data obtained is one of the times when it is the smaller probability. For example, at the .05 level, it is possible the data is one of those 5% of times the data are outside the larger limits (not one of the 95 of 100).

3. Alpha and Beta Errors

hen naming a level in hypothesis testing, such as the .05 or .01 levels, there is a chance the data obtained is one of the rare cases of sampling where $\bar{x}$ does not approximate μ for the level selected. At the .05 level, 95 of 100 times the different $\bar{x}$'s obtained will statistically be the same. What if the sample collected is one of the 5 of 100 times $\bar{x}$ does not approximate μ?

It would seem the thing to do is to set the level high, perhaps at the .01 level. After all, it seems better to be certain 99 of 100 times as opposed to 95 of 100 times. But when setting the level high, risk is reduced in one direction, yet increased in another direction. Alpha (α, Type I) errors are errors made when the hypothesis is true, yet the hypothesis is rejected. Beta (β, Type II) errors occur when the hypothesis is false, and the decision fails to reject it. By *never* accepting the null (there is no difference) hypothesis, an alpha error cannot be made. **Table 11.1** reveals the possibilities in each case.

Table 11.1. Null Hypothesis Decision Chart.

If null hypothesis is true, but rejected, α error.
If null hypothesis is false, but rejected, no error.
If null hypothesis is true, not rejected, no error.
If null hypothesis is false, fail to reject, β error.

No matter what instrument or measure is used, $\bar{x} \neq \mu$. There is always some error in measurement. For example, with I.Q. scores, if the measuring device is precise, no two I.Q.'s are equal. Yet many people are reported as having the same I.Q. By the same analogy, precise instruments of any kind will never have tied scores if the instrument is totally accurate.

When α is small, β is large. By setting higher probability (.05, .01), the null hypothesis should be tested better. Power is 1-β, or the probability of correctly rejecting a false hypothesis. Power can be increased by increasing N. Therefore, set the probability at .05 and use as large an N as possible. When probability is moved to the .01 level, the chance for β errors are decreased. Beta error probability and N required for specific power levels can be found by using z and confidence levels, but the process will not be discussed in this text.

PROBABILITY SAMPLING

Probability sampling may be defined as scientifically created samples that are accepted by statisticians as representing population data. The probability level is defined, that is, .05 level .01 level, etc. The levels are interpreted as the percentage of time the data characteristics will be statistically the same as population data (.05 level , 95% of the time the characteristics are the same, etc.).

1. Random Sampling

In random sampling, each piece of data has an equal chance to be selected. A table of random numbers can be used, although numbers could be drawn by lot in other ways, including a computer program. A simple Basic program that will select 50 random numbers from a universe of 100 follows:

```
50 REM      SELECT 50 RANDOM NUMBERS FROM A
            UNIVERSE OF 100
90 DIM L(100),J(100),R(100)      240 PRINT "-";R(S)
100 FOR J = 1 TO 100             250 R(S) = R(L)
110 R(J) = J                     270 NEXT J
120 NEXT J                       290 PRINT
190:                             300 END
200 FOR J = 1 TO 50
210 L = 100 - J + 1
230 S = INT (RND (1) * L + 1)
```

If a different universe and/or different number of random numbers are desired, *change lines 90, 100, 200 and 210*. For example: select 15 numbers from a universe of 20:

```
90 DIM L (20),J(20),R(20)
100 J = 1 TO 20
200 FOR J = 1 TO 15
210 L=20 - J+1
```

Sampling with replacement means that when a number is used, it is put back into the total *and could conceivably be drawn again*, although the same number would not be used more than one time. The reason is that with a population of known numbers, by replacing the number drawn, it has a chance of being selected again. Even if the number is not selected again, it occupies a place in the total population that may influence which number is selected at any given time.

To use a table of random numbers, which is drawing with replacement, the following procedure may be used:

1.   assign each subject a number.
2.   decide to enter the table at a certain column and row.
3.   decide which digits of the published numbers to use (usually the last two, for less than 100 subjects, or the last three, for 100 - 1000 subjects).
4.   enter the table. Draw the numbers by proceeding down the column to the last number, then go to the next column, etc. An example is shown in **Table 11.2.**

Table 11.2. Numbering Subjects.

1. Susie	6. Autumn	11. Peggy	16. Roberta
2. Anna	7. April	12. Beth	17. Monica
3. Billie	8. June	13. Jean	18. Tracey
4. Carole	9. Bernice	14. Dot	19. Deann
5. Yvonne	10. Joanne	15. Jane	20. Tina

For this sample, 10 of 20 numbers are needed. On a whim, it is decided to enter a table of random numbers at column 3, row 1 of the third thousand (Row 00, Column 11111/01234).

Upon entering **Table 1** in **Appendix B**, the first number listed is 65787. Since only the last two digits will be used, the number 87 is not one of the numbers (1-20) that is needed. The next number underneath is 63918. Since the last two digits, 18, is one of the numbers needed, it is recorded. Continue down the column, then move to the top of the next column, etc. until all 10 numbers needed are found. If a number appears twice, go on to the next number that has not already been used. The final list, in the order found includes the selections shown in **Table 11.3**.

Table 11.3. Subjects Selected From a Table of Random Numbers.

Number	Subject	Number	Subject
18	Tracey	04	Carole
06	Autumn	12	Beth
14	Dot	11	Peggy
02	Anna	13	Jean
15	Jane	20	Tina

Another way of finding 10 of 20 names would be to make a list, put each in a box, and draw 10 names from the box. If sampling with replacement is desired, replace each one drawn and mix them up each time. If a name is drawn again, replace without using it.

2. Stratified Random Sample

In a stratified random sample, the population is divided into subsets, and a *random sample of elements* is taken from each of the subsets. For example students in a small college could be broken down into subsets of class. Then a sample could be obtained of various elements (males or females). **Table 11.4** is an example of the population subsets and elements.

Table 11.4. Subsets and Elements of 881 College Students.

Subset	Element			
	Male (M)	M %	Female (F)	F %
Freshman	213	24%	146	17%
Sophomore	120	14%	94	11%
Junior	102	12%	83	9%
Senior	73	8%	50	5%
Total	508		373	

Note the percentages in **Table 11.4.** They were used to determine the stratified random sample in **Table 11.5** below. Sample size was set at 100, then percentages of each element were determined.

Table 11.5. Stratified Random sample with N = 100.

Subset	Male	Female
Freshman	24	17
Sophomore	14	11
Junior	12	9
Senior	8	5

3. Cluster Sample

In a cluster sample, the population is divided into subsets and a random sample is taken of the subsets. All the elements in the chosen subset are used. With the data in **Table 11.4** above, first randomally determine which of the subsets would be used (freshmen, etc.). Then all of the subset would be sampled; ie, if seniors were the subset, all of the seniors, (73 males and 50 females) would be the target sample.

In the example cited, it is obvious the sample would not truly represent all the data. If the categories (subsets) were of a different character, the sample might be a better representation.

4. Systematic Sample

 systematic sample is one that selects units at a specified interval, such as every 3rd., 10th., etc. If an alphabetical list of names is used, there may be bias, because more people have last names with S than say, A. Therefore, some letters have more representation than others, but perhaps they should have more selections.

NON-PROBABILITY SAMPLES

Non-probability samples are not randomly generated, but they may be useful. *Judgmental* samples use "expert" opinion to find which data is selected. The expert tries to get the best representation possible.

A *quota sample* is made in such a way that the proportion of the elements is in the same proportion as the population. It is similar to a stratified random sample, but the data are not randomly selected.

A *convenient* sample is one that is selected because the data are easy to obtain. Often researchers need to use convenient samples due to time and cost limitations.

SAMPLE SIZE

 ne interesting question is what should the sample size be? The answer is not readily available, because it depends on many factors, such as time, cost, availability, etc. One concept to use is in recalling confidence interval testing. By working backwards through the procedure, an *estimate* of sample size can be obtained.

In confidence interval testing, the concern was, what percent of probability is acceptable for the sample mean? That concern is a judgement call, and there are as many answers as there are people. Using levels that are acceptable to many people, such as the .05 and .01 levels may be a starting point. One basic way to find sample size is shown by equation 24 which is described below.

$$n = \left[\frac{\sigma z}{E} \right]^2$$

(24) *where n = sample size needed;*
σ = estimate of population standard deviation;
z = confidence interval desired;
E = allowed error of sample mean.

The estimate of the population standard deviation can be obtained from a sample pilot study; the confidence level desired (2 tailed test) may be either the .10 (z = 1.64), .05 (z = 1.96), .01 (z = 2.58) levels, or any z level that is appropriate. For confidence levels other than those mentioned, consult a different text. No z tables are provided in this edition.

As an example, assume a pilot study of gross income has found an s of $10,000. A confidence level of .05 is desired, so a 2-tailed z of 1.96 is used. An allowable error of $1,000 is set as the allowable error range. Substituting in equation 24, the sample size is:

$$n = \left[\frac{\acute{o}z}{E} \right]^2$$

$$n = \left[\frac{(10,000)(1.96)}{1,000} \right]^2$$

$$n = \left[\frac{19,600}{1,000} \right]^2$$

$$n = (19.60)^2$$

$$n = 384.16; \; rounded = 384.$$

Using the sample size determined by equation 24 does not influence the outcome of the sample. What the equation is predicting is that the data from 384 clients will result in an error of no more than $1,000 in gross income 95% of the time, provided that the sample s of $10,000 was an accurate estimate of the population σ.

There are two main assumptions in equation 24. They are (a) normality and (b) a random sample. If n is sufficiently large (around 30 may be considered as large), normality is assumed - but not asssured. If sampling procedures are random the other assumption is met.

The allowable error in the example given was determined by judgement of what might be an acceptable error. The error could be stated as a percentage of the sample standard deviation. For example, instead of setting $1,000 as acceptable error, the error could be expressed as a percentage of s. If a 5% error is used, the new error is $10,000*.05 or $500. In the problem mentioned, using a 5% acceptable error would result in a sample size of 1,537 (rounded), more costly and time-consuming in an income survey. But the procedure may be appropriate for some desired samples.

Usually, determining the allowable error is a matter of judgement that can best be estimated by the investigator. For example, if the measure is demographics, what age range age, number in household, income, etc. is a reasonable error? Or if market segment is the criterion, how many business or pleasure travelers would be a reasonable error? Or if travel is used, what length of stay, number in party, etc. are more appropriate? Obviously, it will depend on many factors the researcher will have to consider and evaluate.

It may not be feasible to use a pilot study to obtain an estimated s. If high and low scores are known, an estimate of s can be found by:

$$(25) \qquad s \; (estimate) \; = \; \frac{High\ Score \; - \; Low\ Score}{6}$$

Then s can be used in equation 24 to find the sample size. For example, assume the data used earleir had a high score of $73,210 and the low score was $18,019. Using equation 25, the standard deviation is estimated at $9,918.50. This number could be used in place of $10,000 if a sample s were unknown, but high and low estimates were available.

A correction factor is needed when the sample size is small in relation to the population N. Divide the sample n by the population N, and if the result is 5% or higher, use equation 26 to find the sample n:

$$n_c = \frac{n}{1+\dfrac{n}{N}}$$

(26)   *Where n_c = Corrected sample n*

n = Original sample n

N = Population N

Assuming a population of 7,250 in the sample size computed, n_c is computed as:

384÷7,250 = .053 *or* 5.3%.

Since the percentage ≥ 5%, Find n_c.

$$n_c = \frac{384}{1+\dfrac{384}{7,250}}$$

$$n_c = \frac{384}{1+.053}$$

$$n_c = \frac{384}{1.053}$$

$$n_c = 364.67 = 365$$

As shown by the prior procedure, the new sample size needed is less than the original sample size found. For small sample sizes, such as a pilot study, keep in mind that when using an N less that 31, the data may not be normal. Draw a frequency curve to estimate normality. If the data are not normal, use the sample size with caution.

APPLICATIONS

1. Hypothesis Testing

Use hypothesis testing at convenient levels to determine if data are statistically significant, remembering that there is no way to know if the sample data is in the majority percentage range or not. Convenient levels are usually reported at .05 or .01, but these are judgmental only and are not necessarily right for all occasions. Consider the risk of making a Type I (α) or Type II (β) error in hypothesis testing.

2. Sampling

Probability sampling is defined as having all items with an equal chance of being in the sample. Non-probability sampling is biased in that there is a chance the sample does not represent the population.

In a random sample, subjects are selected randomally for data collection. Random samples are sometimes used in simple random samples, where subjects are listed and selected randomally, such as by computer or by a table of random numbers.

A stratified random sample uses percentages of subsets, characters, etc. that are selected randomally. It would be well to draw a curve of the data to see if the numbers appear to have a normal curve shape. If the curve approximates normality, and the sample size was selected properly, the data probably represent the population, and inferences may be made to the population with caution. If the curve is not normal and/or the sample size not randomly determined, the data should be used and reported with extreme caution as far as inferences beyond the sample collected.

There are many ways to estimate the sample size needed. Different types of samples, such as random versus stratified random samples require slightly different procedures. A text that organizes sampling procedures by type is Ritchie, J.R. Brent; & Goeldner, Charles R. (1987). Travel, tourism and hospitality research. New York: John Wiley & Sons.

Up to now, the text has focused on elementary statistics. Although grouped data could be used for more advanced techniques, it is not advisable to group the data for this purpose, because more accuracy is needed in advanced computations.

It is hoped that by introducing statistics through grouped data, a more thorough appreciation of contemporary technology will inspire the student to examine data relationships and procedures, rather than trying to memorize formulas, which are explained differently in many different texts. Once the concept of deviations squared and variance are learned, the remaining statistical tools are easier to master. Chapter 12 will examine more powerful tools called correlation and regression.

Exercises

1. Using the table of random numbers in Appendix B, starting with page 207, column 11111 (01234), row 00, pick 10 random numbers from a universe of 50.

2. Convert the data in **Table 11.5** to a stratified random sample with an N of 31.

3. Using the data in **Table 11.6** below, find the sample size needed. If appropriate, use n_c to correct the sample size.

Table 11.6. Selected Data for Small Samples.

Problem	s	Allowed Error	Pop. N	CI
a.	1.58	10% s	1,000	.05
b.	11.19	2	1,000	.05
c.	12.84	5% s	5,000	.01
d.	20.52	3.5	250	.01

4. Find the CI ranges at the .05 and .01 levels for problem 4 in the Exercises from Chapter 10.

5. A master planner decides to obtain client preferences in each neighborhood of a city. She decides to survey people based on age, because other variables are not practical for a sample base. She will distribute surveys based on a low age of 12 and a high age of 93. There is not time to do a pilot study to obtain the standard deviation. She wants to consider an error in age of 1 year, 2 years and 5 years. Based on these error ranges, what sample sizes are needed at the .05 level?

6.   Use **Table 11.7** to find sample sizes needed at the .05 level.

Table 11.7. Fictitious Data.

Sample Type	Low	High	Error Accepted
a. Client Income	$25,000	$85,000	$500; $1,000
b. Travel Agency (Age)	16	72	5 years
c. Business Traveler Income	$45,000	$245,000	$1,000; $2,000

Chapter 12

Correlation

1. Introduction

orrelation is defined as the *relationship one set of data has with a corresponding set of data*. Specifically, it is the ratio of dependent to independent sums of squared deviations from the mean. Sums of squared deviations from the mean are similar to the sums of deviations squared determined as a first step in standard deviation. In fact, the *independent* deviations are found in the very same way.

The independent deviations (Σx^2, Σy^2) refer to *how much the two sets of data differ from each other*. To find the independent deviations, find Σx^2 for each set of data. Since there are two sets of data, one set of scores are labeled X scores, and Σx^2 is found in the usual way. The other set of data is labeled as Y scores, and the sum of deviations squared is found in the same way, but are labeled Σy^2. The two values, Σx^2 and Σy^2 are multiplied together and the square root of this value is the denominator for the correlation ratio. The X data are independent variables (predictors), while the Y data are dependent variables (what is predicted).

The *dependent* deviations (Σxy) refers to *how much the scores vary together*. To obtain this value, a new column *equivalent* to squared data called the XY column is created. Each X score is multiplied by the **corresponding** Y score (for similarity, a number times itself is squaring). The Σxy is then obtained in a way that is similar to finding the sum of deviations squared for X data. Column entries for X and Y are used. Each entry in the X column is multiplied by each entry in Y column and summed to obtain the ΣXY data.

The definitional formula for obtaining the correlation coefficient (r) is as follows:

(27)
$$r = \frac{\Sigma xy}{\sqrt{\Sigma x^2 \Sigma y^2}}$$

In determining the numerator and denominator for equation 27, the sum of deviations squared for y and xy are similar to obtaining the sum of squares for x. Using equation 27, the procedure for computing r is:

1. $\Sigma x^2 = \Sigma X^2 - \dfrac{(\Sigma X)^2}{N}$

2. $\Sigma y^2 = \Sigma Y^2 - \dfrac{(\Sigma Y)^2}{N}$

3. $\Sigma xy = \Sigma XY - \dfrac{(\Sigma X)(\Sigma Y)}{N}$

4. $r = \dfrac{\Sigma xy}{\sqrt{\Sigma x^2 \Sigma y^2}}$

The data in **Table 12.1** below will be used as an example.

Table 12.1. Hand Weights Used in Walking.

	Pounds Trial 1		Pounds Trial 2		
Subject	X	X²	Y	Y²	XY
1	1	1	5	25	5
2	2	4	1	1	2
3	3	9	4	16	12
4	4	16	2	4	8
5	5	25	3	9	15
Σ	15	55	15	55	42

Example: A simple example of correlation where inferences cannot be made as to cause and effect follows. Five subjects were asked to pick hand weights to use in walking. After walking one mile each, they were asked to choose weights again. Was there a relationship between practice walks and the choice of weight in pounds from one walk to the next? **Table 12.1** on the previous page presents the data obtained in tabular form.

Following steps given earlier, the solution is:

1. $\Sigma x^2 = \Sigma X^2 - \dfrac{(\Sigma X)^2}{N} = 55 - \dfrac{(15)^2}{5} = 55 - 45 = 10$

2. $\Sigma y^2 = \Sigma Y^2 - \dfrac{(\Sigma Y)^2}{N} = 55 - \dfrac{(15)^2}{5} = 55 - 45 = 10$

3. $\Sigma xy = \Sigma XY - \dfrac{(\Sigma X)(\Sigma Y)}{N} = 42 - \dfrac{(15)(15)}{5} = -3$

4. $r = \dfrac{\Sigma xy}{\sqrt{\Sigma x^2 \Sigma y^2}} = -\dfrac{3}{\sqrt{10*10}} = -\dfrac{3}{10} = -.30$

2. Discussion

To interpret this statistic, find **Table 2**, Values of the Correlation Coefficient for Different Levels of Significance, in **Appendix B**. For a one-tail test, enter the column marked n or df (degrees of freedom) with a value equal to N-2 (N= number of *paired* X, Y values), which in this case is 5-2= 3. Then select the level of significance desired (.05, .01., etc.) and *double* it (.05 = .01 column; .01 = .02 column; etc.).

For example, for p = .05, go down the n column to 3, come across the row where p =.01, and read the significance level right off the chart. The value in the chart is the r value needed for significance at the level selected. With this data, an r of .805 or more is needed for significance at the .05 (95% of the time) level. Since the r obtained is only -.30, the r is not significant at the .05 level.

Two-Tailed Test. So far, the discussion has centered around a one-sided test. That is, when there is reason to suspect that a correlation exists in a specific direction, then a one-sided test may be used. However, if there is no preliminary indication, it is more appropriate to use a two-sided test. In other words, the relationship could be either plus or minus, and all that is necessary is to find the r value needed is to locate the desired p column and read the significance value directly from the table.

To illustrate, in the present problem, a two-tailed test at the .05 level reveals that an r of .8783 is needed. To find this value enter the table with 3 df (n), move to the column with a heading of .05, then read the table value value right off the chart, which is .8783. For a two-tailed test, an r of .8783 or higher is needed. Notice that when moving from a one to a two-tailed test, the *r value needed for significance is increased.*

Interpretation. To further understand r values, a negative (-) value means there is an *inverse* relationship, while a positive (+) value indicates a *direct* relationship. Also, it is generally understood in the literature that a plus or minus r of 0 to .30 usually means there is **no** relationship, .31-.60 shows **some** relationship, while above .61 indicates a **high** relationship. The highest possible r is ±1.00, which would be a perfect correlation.

In the problem above, **Table 2** in Appendix B shows the r of -.30 is not significant at the .05 level, and using the general guidelines (0-.30 = none) there is probably no relationship to the data. If any minor relationship exists it is inverse; selections made after the second trial were in a direction opposite to the selection made the first time. That is, if a person chose a heavier weight the first time, they were more inclined to choose a lighter weight the second time, and vice-versa. A glance at the data will verify this conclusion. However, the data showed no correlation at all, and conclusions of anything but this interpretation should be disregarded.

Even though the data were not significant, insight can still be gained from examination of **Table 12.1**. Notice that with 30 df in a one-tailed test, an r of .30 would be significant at the .05 level.

While discussing grouped data, it was mentioned that increasing N would decrease $s_{\bar{x}}$. **Table 12.1** reveals that increasing N would increase significance, if the r were to remain the same. Prior to gathering data, it might be helpful to increase N to 31 or more scores. That way, significance or non-significance due to a low N may be reduced, along with errors that are associated with the mean. In the problem at hand, increasing N to 31 would result in a significant relationship if the r obtained remained the same.

3. Direct Determination of Sums of Deviations Squared

The data in **Table 12.1** can be used to show the Σx^2 can be computed using deviation scores from the mean. Figure 12.1 is derived from **Table 12.1** and explains the process by deriving deviations (x) and squaring same.

Subject	X	Trial 1 X Data — Mean	x	x²	Y	Trial 2 Y Data — Mean	y	y²	xy
1	1	3	-2	4	5	3	2	4	-4
2	2	3	-1	1	1	3	-2	4	2
3	3	3	0	0	4	3	1	1	0
4	4	3	1	1	2	3	-1	1	-1
5	5	3	2	4	3	3	0	0	0
		Σ	0	10			0	10	-3

Figure 12.1. Sum of Deviations From the Mean.

Notice that the Σx^2 and the Σy^2 are 10, the same value found in the original computation. By examining the data closely, it is obvious how the deviations (x) and sum of deviations squared (Σx^2) are determined. The Σx and Σy deviations are found by subtracting the mean from each individual X or Y score. These columns are then squared and *added* to obtain Σx^2 and Σy^2. By the same token, each X deviation (x) is multiplied by each Y deviation (y) to result in the xy column, *which is then summed* to become Σxy. Study the above data carefully, because it is important to understand what these x, y and xy values represent before proceeding further.

4. Line Slope

orrelation can also be found by taking the square root of the product of the slopes of each line. The statistic **byx** is the *tendency of Y to meet X* and is found by dividing Σxy by Σx². The slope **bxy** is the *tendency of X to meet Y* and is found by dividing Σxy by Σy². Equation 28 on the following page shows these procedures mathematically.

$$\text{(28)} \qquad 1.\ byx = \frac{\Sigma xy}{\Sigma x^2} \qquad 2.\ bxy = \frac{\Sigma xy}{\Sigma y^2}$$

In the problem regarding use of hand weights:

$$1.\ byx = \frac{-3}{10} \qquad 2.\ bxy = \frac{-3}{10}$$

$$byx = -.3 \qquad bxy = -.3$$

This example is interesting, because there is a negative byx and bxy. The correlation coefficient is a pure number, so the following procedure applies:

$$r = \sqrt{byx*bxy}$$

$$r = \sqrt{(-.3)(-.3)}$$

$$r = \sqrt{-.3}*\sqrt{-.3}$$

$$r = -.3$$

The case above where two negative numbers are multiplied under the radical is the only time byx and bxy will need to be broken into two separate radicals and multiplied together for a negative solution. It is only logical that when Σxy is negative, both byx and bxy will be negative, since Σxy is the numerator of both byx and bxy. And a negative Σxy will result in a negative correlation ratio. Therefore, whenever Σxy is negative, the correlation ratio must be negative, and when this ratio is determined in the manner described above, the result will have to be listed as negative. Another way to think of this situation is to realize the correlation ratio is a number with an absolute value that takes the sign of the dependent variable, Σxy.

5. Scatter Plot

A visual look at what is occurring is to draw a scatter plot. Plot the X scores against the Y scores and the relationship, if any, can be seen as a straight line joining all points or one that best describes the general slope and direction of these points. It may be easier to plot these scores by making a table with values of X and Y, then plot the X score against the corresponding Y score. **Table 12.2** shows corresponding X and Y values, and Figure 12.2 plots the paired scores.

Table 12.2. Fictitious Paired Data.

X	1	2	3	4	5
Y	5	1	4	2	3

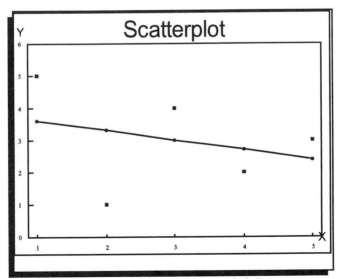

Figure 12.2. Scatterplot of Table 12.2 Data.

If there is a line of best fit, a straight line could be drawn that will have an average distance away from points above and below the line.

With the data in question, the line runs almost parallel to the X axis, and this analysis further confirms the fact there is no significant correlation in choosing hand weights between the two trials.

Further computation is needed to find where byx intercepts the Y axis. To find this value, the following method is given:

(29)

$$a_{yx} = \overline{Y} - b(\overline{X})$$

$Where$: a_{yx} = Y axis intercept

$\overline{Y}$ = mean of Y

b = byx

$\overline{X}$ = mean of X

Substituting values found earlier in the example for equation 28:

$$a_{yx} = 3 - (-.3) * 3$$

$$a_{yx} = 3.9$$

6. Line of Best Fit

In order to draw a line of best fit, it is necessary to draw a scatter plot. If there is a perfect correlation ($r = +1.00$ or -1.00), a straight line can be drawn through *all* the points. If a straight line cannot be drawn through all points, then a line that is equal distance above and below each X and Y point is drawn. The resulting line is called a regression line, and in this case is the regression line Y on X. The point where it intercepts the Y axis can be found by a_{yx}, and connected to other plotted points found by substituting known values in equation 28. To reinforce by repetition the procedure, the following example is given.

Example: Seven children in a fitness center were asked to do as many push-ups as they could in ten seconds. They were then put into a three week training program to increase upper body strength. At the end of the three weeks, they were asked to perform the test again. Was there a statistical *relationship* between the number of push-ups during the three week period (see **Table 12.3**)?

Table 12.3. Push-Ups in Ten Seconds.

	Trial 1		Trial 2		
Subject	X	X^2	Y	Y^2	XY
1	2	4	4	16	8
2	4	16	5	25	20
3	5	25	8	64	40
4	7	49	8	64	56
5	9	81	7	49	63
6	8	64	9	81	72
7	3	9	1	1	3
Σ	38	248	42	300	262

To solve the problem, first determine Σx^2, Σy^2, Σxy, r, byx, a_{yx} and the regression equation [Y=a+b(X)]. An example follows.

Before drawing the scatterplot, it may help to develop a table of X versus Y plots. **Table 12.4** on the following page is an example of such a table.

Table 12.4. X, Y Plots From Table 12.3.

X	2	4	5	7	9	8	3
Y	4	5	8	8	7	9	1

In addition, in order to draw the regression line, it will be necessary to solve the prediction equation for a straight line in order to predict X and Y values for a line of best fit. Equation 30 follows below:

$$Y = a+b(X)$$

Where: Y = Y axis value

(30)
$$a = a_{yx}$$

$$b = byx$$

$$X = X \text{ value}$$

1. $\Sigma x^2 = \Sigma X^2 - \dfrac{(\Sigma X)^2}{N} = 248 - \dfrac{38^2}{7} = 41.72$

2. $\Sigma y^2 = \Sigma Y^2 - \dfrac{(\Sigma Y)^2}{N} = 300 - \dfrac{42^2}{7} = 48.00$

3. $\Sigma xy = \Sigma XY - \dfrac{(\Sigma X)(\Sigma Y)}{N} = 262 - \dfrac{(38)(42)}{7} = 34$

4. $r = \dfrac{\Sigma xy}{\sqrt{\Sigma x^2 \Sigma y^2}} = \dfrac{34}{\sqrt{(41.72)(48.00)}} = .759$

5. $b_{yx} = \dfrac{\Sigma xy}{\Sigma x^2} = \dfrac{34}{41.72} = .81$

6. $a_{yx} = \overline{Y} - b(\overline{X}) = 6.00 - .81(5.43) = 1.60$

To solve equation 30 with the data above:

$Y = a+b(X)$

$Y = 1.60 + .81(1)$

$Y = 2.41$

Interpreting the equation, Y is 2.41 when X is 1. By using different values for X, **Table 12.5** below will be used to plot the *line of best fit* on the scatterplot of Figure 12.3.

Table 12.5. Rounded X,Y Plots for Y=a+b(X).

X	2	4	5	7	9	8	3
Y	3.2	4.8	5.7	7.3	8.9	8.1	4.0

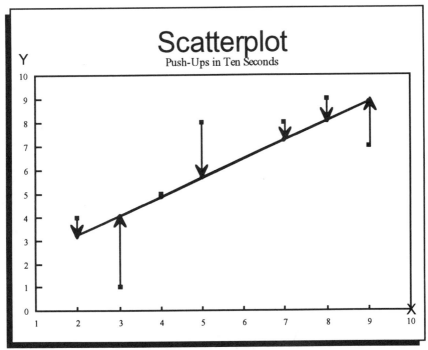

Figure 12.3. Scatterplot From **Table 12.5**.

There is no straight line that will go through *all* points on the scatter plot. However, it is possible to construct a regression line that will have a total distance the *smallest sum of all the squared deviations* from each point directly to the line of best fit. Also notice the Y axis was constructed close to .67 of the X axis as when drawing all graphs from Chapter 7.

7. Simplifying Columns

If one or both of the columns has large numbers, it may be desirable to subtract a constant from all the numbers in a column because the resulting smaller numbers are usually easier to work with and may reduce chances for error. X scores in **Table 12.6** refer to a the number of repeat mailings of a program brochure compared to the number of clients (Y scores) that actually signed up for the programs.

Table 12.6. Number of Program Brochures mailed.

X	X^2	Y	Y^2	XY
1	1	60	3,600	60
2	4	56	3,136	112
3	9	59	3,481	177
4	16	57	3,249	228
5	25	58	3,364	290
Σ 15	55	290	16,830	867

Solution:

$$1.\ \Sigma x^2 = 55 - \frac{15^2}{5} = 10$$

$$2.\ \Sigma y^2 = 16,830 - \frac{290^2}{5} = 10$$

$$3.\ \Sigma xy = 16,830 - \frac{(15)(290)}{5} = -3$$

$$4.\ r = \frac{\Sigma xy}{\sqrt{\Sigma x^2 \Sigma y^2}} = \frac{-3}{\sqrt{10*10}} = -.3$$

Subtracting a constant is shown in **Table 12.7**.

Table 12.7. Subtracting a Constant From Table 12.6.

X	X²	Y	New Y = Y-55	Y²	XY
1	1	60	5	25	5
2	4	56	1	1	2
3	9	59	4	16	12
4	16	57	2	4	8
5	25	58	3	9	15
Σ 15	55		15	55	42

The solution using **Table 12.7** is:

$$1. \ \Sigma x^2 = 55 - \frac{15^2}{5} = 10$$

$$2. \ \Sigma y^2 = 55 - \frac{15^2}{5} = 10$$

$$3. \Sigma xy = 42 - \frac{(15)(15)}{5} = -3$$

$$4. \ r = \frac{-3}{\sqrt{10*10}} = -.3$$

In both cases, the same Σx^2, Σy^2, Σxy and r are obtained, even though the constant number of 55 was subtracted from each score in the Y data. Certainly, when there are only a few scores, this procedure would not be necessary, and if using a computer program it may not be needed. But for hand calculations with a lot of large numbers, it may be easier to subtract a constant in order to work with smaller numbers. It may also help reduce errors caused by copying large numbers by hand.

8. Pearson Product Moment

The *Pearson* correlation ratio is found the same way as the procedure using deviations. However, it is usually expressed by the following formula:

$$(31) \quad r = \frac{N\left[\Sigma XY - (\Sigma X)(\Sigma Y)\right]}{\sqrt{\left[N\Sigma X^2 - (\Sigma X)^2\right]\left[N\Sigma Y^2 - (\Sigma Y)^2\right]}}$$

The serious student will notice this expression is nothing more than the definitional deviations reduced to the lowest algebraic terms.

9. Multiple Regression

Multiple regression can be used when the data is at least interval in nature. Sometimes there are more than two sets of data. In this case, correlation ratios can be found in the same manner as other correlations by using Y as one set of scores, and finding r for each of the other sets of data in turn, one by one. For example, if there are four sets of data, the Y data is the criterion (the standard against which all other variables are judged, such as time, distance, volume, etc.). The Y values are also known as the dependent variables - data that is known now but for which a future prediction is needed.

The other data may be labeled as X_1, X_2 and X_3. Then a correlation ratio can be determined between all of the variables in question. Also, intercorrelations can be found (X_1 and X_2, X_1 and X_3, and X_2 and X_3).

Scatter plots can be drawn as before, using each set of data. Sometimes the scatter plots can all be drawn on the same graph. Data from one set of samples may not be a reliable predictor for other samples.

A prediction equation can be found for multiple correlations, and is known as a multiple regression equation. It is found the same way as when two sets of data are used, but additional factors are weighted into the equation, such as:

$Y = a + b(X_1) + b(X_2) + b(X_3). \ . \ . \ +b(X_j)$

Where: Y = dependent variable

    a = constant (a_{yx})

    b = byx for each set of X data (independent variable)

    $X_1, X_2, X_3, \ . \ . \ .X_j$ = each set of X data, 1 through j

(32)

One way to approach a multiple regression problem is to use a computer program to calculate correlations and intercorrelations, and test each set of scores for significance in the usual way. Ratios found to be significant can then be used as the remaining X scores to determine the multiple regression equation. A technique known as partial correlation $(r_{12 \cdot 3})$ can help determine which coefficients are more meaningful.

There is an error range associated with the variance in correlation techniques that is expressed by R^2. This statistic refers to the variance accounted for in the sample. For example, and r of .800 accounts for only 64% $(.800^2)$ of the variance in two sets of scores (see page 201).

10. Standard Error of Estimate

Another error, known as standard error of estimate, is found by:

(33) $S_{y*x} = s\sqrt{1.00 - r^2}$ Where:

    S_{y*x} = standard error of estimate

    s = standard deviation of dependent variable(Y)

    r = reliability coefficient of the scores

The computed $s_{y.x}$ is the standard error with a probability of occurrence 68.26% of the time. Applied to a single measured score, it is used the same way as confidence interval testing. Higher probability levels such as .05 (1.96) or .01 (2.58) can be multipled by the $s_{y.x}$ and when added and subtracted from a score yields an error range the same as in confidence interval testing.

11. Standard Error of Measurement

 he standard error of measurement is a procedure similar to standard error of estimate and is found by equation 34 below. The SE_M is determined, applied to a score, and used the same way as the standard error of estimate.

$$SE_M = s\sqrt{1-r}$$

(34) *Where: SE_M = standard error of measurement*

s = standard deviation of scores on a test

r = reliability coefficient of the test

12. Assumptions of r

 orrelation techniques and regression equations are powerful tools that are often used in analyzing human performance. Although cause and effect is not established or ruled out, relationships can be examined, as long as an underlying theory is plausable and data characteristics are honored. The following assumptions are made about the data when using correlation procedures discussed in this chapter:

1. bivariate random variables.
2. continuous data (see chapter 6).

3.   normality (a normal curve distribution).
4.   interval data or higher (see chapter 6).
5.   linearity (the data represent a straight line).
6.   homoscedasity (equal varainces).

To the extent that these assumptions are or are not met, the correlation/regression techniques are or are not appropriate.

In regard to linearity, it is possible for two sets of data to be highly related yet not be linear. Other correlation techniques exist for such data, including biserial, point biserial, tetrachoric, and phi. Space will not permit a discussion of these methods, but Chapter 13 is devoted to rank correlation techniques.

APPLICATIONS

It is important to realize that correlations, or any realtionship between variables, may be *experimental* or *relationship* oriented. In experimental studies, factors with controlled amounts of the variable under study are randomly assigned treatments while observing the effect on a different variable. For example, if an experimental study of the effect of income on client choice were desired, subjects would be given varying amounts of money and the effect on client choce would be observed.

If the amount of income were determined for each subject and compared withclient choice, the study would be relationship oriented. No random placement of income with different income amounts are used in this case. Connections between variables are chance, and the income level is not initally known.

Many of the studies in leisure services, sport management and travel are difficult to control experimentally. Income distribution is not randomly assigned - people have amounts they start with. By the same token, demographics such as age, sex, etc. are not metered out, then randomly assigned. These parameters can be measured and categorized, then clients can be randomly assigned, but it is not quite the same as randomly assigning measured amounts of fertilizer on corn yield, or randomly assigning different amounts of a chemical to study cell changes. In correlational studies, random amounts of variables under study already exist, but they exist by chance, *not through random assignment*. Regardless of the form the study may take, it can still be useful for learning about relationships that variables may have. An example of a *correlational analysis* follows.

Example: Average snow depth in 8 years were measured and compared with the total number of paid skiers at a ski area. Was there a linear relationship? What was the correlation ratio? Interpret the r ratio. Using a prediction equation, what would the expected number of skiers be for a snow depth of 50 inches? If a year had 30,000 paid skiers, what expected average snow depth would the manager expect to have measured? **Table 12.8** lists the data.

Table 12.8. Fictitious Snow Depth and Number of Skiers.

	'81	'82	'83	'84	'85	'86	'87	'88
Avg. Depth (inches)	41	38	47	53	32	44	35	39
Skier Count (1,000)	25	22	26	20	12	26	20	27

To solve this problem, determine a correlation ratio. Check for significance. Draw a scatter plot. Determine byx and bxy, and solve the prediction equation $Y = a + b(X)$. **Table 12.9** lists the data.

Table 12.9. Snow Depth and Number of Skiers for 8 Years.

X	X^2	Y	Y^2	XY
41	1,681	25	625	1,025
38	1,444	22	484	836
47	2,209	26	676	1,222
53	2,809	20	400	1,060
32	1,024	12	144	384
44	1,936	26	676	1,144
35	1,225	20	400	700
39	1,521	27	729	1,053
Σ 329	13,849	178	4,134	7,424

Problem solution:

1. $\bar{X} = \dfrac{329}{8} = 41.13;\ \ \bar{Y} + \dfrac{178}{8} = 22.25$

2. $\Sigma x^2 = \Sigma X^2 - \dfrac{(\Sigma X)^2}{N} = 13{,}849 - \dfrac{329^2}{8} = 318.88$

3. $\Sigma y^2 = \Sigma Y^2 - \dfrac{(\Sigma Y)^2}{N} = 4{,}134 - \dfrac{178^2}{8} = 173.50$

4. $\Sigma xy = \Sigma XY - \dfrac{(\Sigma X)(\Sigma Y)}{N} = 7{,}424 - \dfrac{(329)(178)}{8} = 103.75$

5. $r = \dfrac{\Sigma xy}{\sqrt{\Sigma x^2 \Sigma y^2}} = \dfrac{103.75}{\sqrt{(318.88)(173.50)}} = \dfrac{103.75}{235.21} = .441$

With 8 - 2 = 6 df, the table value needed for significance at the .05 level is .6215. Since the r value is only .441, the data were not significant at the .05 level.

However, using the general interpretation guidelines, there may be some correlation (.31-.60 = some). Draw the scatter plot to determine if a line of best fit is possible (see Figure 12.4 on the next page).

To answer the other questions, it is necessary to first determine byx and a_{yx}, then solve tthe prediction equation.

6. $byx = \dfrac{\Sigma xy}{\Sigma x^2} = \dfrac{103.75}{318.88} = .325$

7. $a_{yx} = \bar{Y} - b(\bar{X}) = 22.25 - .325(41.13) = 8.88$

8. *Using 6 and 7 above, the prediction equation is:*

   $Y = 8.88 + .325(X)$

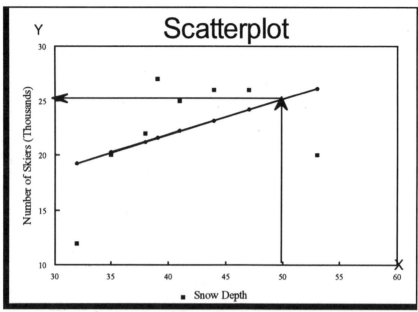

Figure 12.4. Scatterplot of Snow Depth and Number of Skiers.

Now the prediction equation can be used to solve the rest of the problem. To determine the number of skiers expected with a snow depth of 50 inches,

Y = 8.88 + .325(50) = 8.88 + 16.25 = 25.13.

Therefore, it is predicted that a snow depth of 50 inches would have 2,513 paid skiers. By inspection of the scatter plot arrows in Figure 12.4 this value looks fine.

Finally, what snow depth would be expected for an attendance of 3,000 paid skiers? Again:

Y = 8.88 + .325(X).   Since Y (30) is known,

30 = 8.88 + .325X

21.12 = .325X

64.98 = X

A glance at the scatter plot shows this value appears to be about right. The 8.88 may be thought of as a correction factor needed to equalize the scales. Snow depth is measured in inches, whereas number of skiers is a different measure. However, mathematically, 8.88 is the point where the line of best fit intercepts the Y axis when X is 0. The .325 means that every inch increase in snow depth will result in about 325 more paid skiers.

Prediction equations are usually unreliable when applied to groups other than the ones from which the prediction was made. A technique called cross validation can be used to check the accuracy of the prediction.

Variance Accounted For. A value that can help explain the variance accounted for is known as R^2. R^2 is a ratio of the independent to dependent variables, and will lie between 0 and 1. the closer the value is to 1, the closer relationship there is between independent and dependent variables. In the problem discussed, r was .441, so R^2 (r^2) is .194. The low relationship is a reflection of the r which was determined as non-significant at the .05 level initally.

There are many other methods of correlation techniques that will not be examined in this text. Chapter 13 will examine non-parametric methods of correlations by ranking data, as well as the Kuder-Richardson and split halves techniques of finding internal consistency reliability.

Exercises

1. Given the following data:

Table 12.10. Hypothetical Millions of Vacation Person-Trips.

Month	1986	1987
January	20.20	20.30
February	21.54	20.76
March	20.50	20.85
April	21.06	21.00
May	20.90	21.18
June	21.80	21.23
July	21.40	21.23
August	21.54	21.27
September	21.80	21.49
October	21.40	21.52
November	21.44	21.62
December	21.40	21.71

a. Find r, byx, bxy, prediction equation.
b. Draw a scatter plot of the data.
c. Explain what the prediction equation means.
d. Interpret r at the .05 level with a one-tailed test.
e. What is R^2?

2.   Listed below are attendance figures in 22 programs.

Table 12.11. Attendance in 1987 & 1988.

Program	1987 (X)	1988 (Y)	Program	1987 (X)	1988 (Y)
1	37	66	12	42	67
2	51	68	13	42	55
3	28	45	14	39	56
4	37	67	15	29	55
5	42	55	16	48	42
6	24	55	17	32	67
7	31	50	18	35	91
8	32	65	19	23	57
9	38	61	20	32	33
10	47	43	21	25	59
11	33	62	22	34	63

a. Find r, byx, bxy, prediction equation.
b. Draw a scatter plot of the data.
c. Explain what the prediction equation means.
d. Interpret r.

A chamber of conmerce in a small town recorded the following office activities (modified from Colbert, 1970).

Table 12.12. Office Activities.

	Office	Walk-Ins	Mail	Answered	
Week	Local	Tourists	Relocate	Tourists	Other
1	93	97	97	88	94
2	83	96	103	99	101
3	97	100	109	108	105
4	102	100	99	96	99
5	141	88	101	90	83
6	95	78	87	90	91
7	77	75	73	75	78
8	97	94	96	91	87
9	115	90	86	88	89
10	127	131	109	84	118
11	88	94	91	86	91
12	68	76	81	91	77
13	101	114	83	85	78
14	82	80	89	81	78
15	104	91	102	107	110

3.   Using **Table 12.12**, with local office walk-ins as X and other mail answered as Y, determine the following statistics:

a. r, byx, bxy, prediction equation.
b. draw the scatter plot.
c. explain what the prediction equation means.
d. Interpret r with a one-tailed test.

4. Using the local office walk-ins as X scores in **Table 12.12** and tourist mail answered as Y scores, determine the following statistics:

   a. r, byx, bxy, prediction equation.
   b. draw the scatter plot.
   c. explain what the prediction equation means.
   d. interpret r with a one-tailed test at the .05 level.

5. Graph a scatter plot for local walk-ins (X) and mail answered (relocate) (Y) scores in **Table 12.12**.

A questionnaire was declared valid by a panel of judges. In order to check reliability, the questionnaire was given to 31 people on a test-retest procedure. **Table 12.13** summarizes the results.

Table 12.13. Personal Data Questionnaire Test-Retest.

Question Number	Number Answering First Test	Number Answering Test 2 The Same
1	31	31
2	31	31
3	29	29
4	31	31
5	31	25
6	31	28
7	31	29
8	31	30
9	17	17
10	31	31
11	3	3
12	30	29

6. Using **Table 12.13**, determine reliability by finding and interpreting the correlation coefficient. How do you interpret the whole questionnaire? Each question?

7. Using a computer program, make a scatter plot for office walk-ins (tourists) (X) and mail answered (tourists) (Y) in **Table 12.12**.

8. Determine a multiple regression equation for mail answered (relocate) (Y or criterion), walk-ins (local) and mail answered (tourists) in **Table 12.12**.

Given the following information:

Table 12.14. Chamber of Commerce Walk-Ins and Hotel Occupancy Rate.

	Walk-Ins (X)	Occupancy Rate (Y)
January	37	70%
February	61	65%
March	81	67%
April	97	68%
May	88	67%
June	156	80%
July	203	90%
August	254	92%
September	145	75%
October	131	65%
November	63	60%
December	31	75%

9.   Using **Table 12.14**, find:

a. r, byx, bxy, prediction equation.
b. draw the scatter plot.
c. explain what the prediction equation means.
d. interpret r with a one-tailed test at the .05 level.

Table 12.15. Travel Sales and GNP During the 1980's[1].

Year	Travel Industry Sales	Gross National Product (GNP)
1980	10.8	8.9
1981	11.0	11.7
1982	4.8	3.7
1983	7.9	7.6
1984	7.6	10.8
1985	4.8	6.4
1986	8.8	5.4
1987	10.3	6.7
1988	8.9	7.9
1989	7.2	6.7

[1]U.S. Travel Data Center. (1990). 1990 Outlook for travel and tourism. Washington: U.S. Travel Data Center.

10.   Using **Table 12.15**, fnd:

a. r, byx, bxy, prediction equation.
b. draw the scatter plot.
c. explain what the prediction equation means.
d. interpret r with a one-tailed test at the .05 level.

REFERENCES

Colbert, C.C. (1970). A comparative study of a jogging program versus a running program for improvement of cardiovascular fitness. Unpublished master's thesis, The University of Georgia, Athens.

Chapter 13

Rank Correlation

1. Spearman Rho

The *Pearson Product-Moment* and definitional methods of correlation assume the data is at least interval. When the data are ordinal, or rankable only, then a different method of correlation must be done. One way to solve this problem is to use the *rank-difference* (Spearman-Rho) procedure.

In the rank-difference method, the data is obtained and the raw scores are converted to ranks. A difference between ranks (d) is found, squared, summed and used as follows:

$$P = 1 - \frac{6\Sigma d^2}{N(N^2 - 1)}$$

(35) *Where*: P = *rank correlation coefficient*

Σd^2 = *sum of rank deviations squared*

N = *number of paired scores*

Example: The Cowell Personal Distance Inventory was given to clients at the beginning of a workshop in recreational games. After playing social games during a period of 12 weeks, the inventory was administered again. Was there a relationship in personal distance scores between the first and last values as shown by the Cowell test? **Table 13.1** lists the scores of each subject. (See Appendix A for a sample personal distance inventory. Personal distance is defined as how close a person is willing to accept another person as a brother or sister. A low score is closer, hence a higher rank.). A high relationship between the data may indicate the lower a person's score was initally, the more potential that person has for a decreased social distance score later.

Table 13.1. Cowell Personal Distance Scores of 21 Students.

S^1	Test 1	Rank 1	Test 2	Rank 2	d	d^2
1	90	8	87	6	2	4.00
2	88	7	90	8	-1	1.00
3	96	10	98	17.5	-7.5	56.25
4	102	17.5	95	13.5	4	16.00
5	107	20	101	19.5	0.5	0.25
6	82	3	75	1.5	1.5	2.25
7	77	2	83	5	-3	9.00
8	86	5	89	7	-2	4.00
9	95	9	92	10	-1	1.00
10	87	6	82	4	2	4.00
11	102	17.5	93	11.5	6	36.00
12	97	12	96	15.5	-3.5	12.25
13	101	15.5	91	9	6.5	42.25
14	84	4	79	3	1	1.00
15	98	14	96	15.5	-1.5	2.25
16	97	12	98	17.5	-5.5	30.25
17	108	21	95	13.5	7.5	56.25
18	76	1	75	1.5	-0.5	0.25
19	106	19	101	19.5	-0.5	0.25
20	97	12	93	11.5	0.5	0.25
21	101	15.5	103	21	-5.5	32.25
Σ	1,977		1,912		0	309.00

^{1}S = Subject. A solution follows on the next page.

$$P = 1 - \frac{6(\Sigma d^2)}{N(N^2-1)}$$

$$P = 1 - \frac{6(309)}{21(21^2-1)}$$

$$P = 1 - \frac{1,854}{9,240}$$

$$P = 0.799$$

Referring to Table 3 in Appendix B, with N-2 df (19), a value of .399 is needed for significance at the .05 level (19 df is not listed. Choose the closest df that is a more stringent test, ie., 18). Since an r of .799 was obtained, the data is significant beyond the .05 level, meaning the sample scores are different 95 of 100 times in social distance. Inspection of the data shows the scores in test 2 were lower, meaning social distance decreased during the 12 weeks.

The rank-difference method is a powerful tool because there are fewer assumptions when using non-parametric data. P *approximates* the Pearson method. A scatter plot could be drawn of the data in the same way as with raw scores. However, time, golf scores or other measures such as social distance that result in low scores representing a higher value should be recorded with low values in higher positions.

2. Tied Ranks (r_s)

When there are a large number of tied ranks, the ties may affect the result. Therefore, a different equation (36) may be a better test.

(36)
$$r_s = \frac{\Sigma x^2 + \Sigma y^2 - \Sigma d^2}{2\sqrt{\Sigma x^2 \Sigma y^2}}$$

Where: r_s = *rank correlation corrected for ties*

AND (*see next page*)

(continued from previous page)

$$\text{(36)} \quad \Sigma x^2(\Sigma y^2) = \left[\frac{N^3-N}{12}\right] - \Sigma T_x$$

$$\Sigma T_x = \frac{(\text{\# ranks tied})^3 + same + (...N_j \ ranks \ tied)}{12}$$

The following ties are noted in **Table 13.2**:

Table 13.2. Tied Ranks In **Table 13.1**.

Test 1 (X)		Test 2 (Y)	
Number Tied	Data	Number Tied	Data
2	102	2	101
2	101	2	98
3	97	2	96
		2	95
		2	93
		2	75

Procedure:

$$r_s = \frac{\Sigma x^2 + \Sigma y^2 - \Sigma d^2}{2\sqrt{\Sigma x^2 \Sigma y^2}}$$

1. Find Σx^2.

2. Find Σy^2.

3. Substitute values, including d^2, into equation 36.

1. $\Sigma x^2 = \dfrac{N^3-N}{12} - \Sigma T_x$

$\Sigma x^2 = \dfrac{21^3-21}{12} - \left[\dfrac{2^3-2}{12} + \dfrac{2^3-2}{12} + \dfrac{3^3-3}{12}\right]$

$\Sigma x^2 = \dfrac{9,261-21}{12} - \left[\dfrac{8-2}{12} + \dfrac{8-2}{12} + \dfrac{27-3}{12}\right]$

$\Sigma x^2 = \dfrac{9,240}{12} - \left[\dfrac{6}{12} + \dfrac{6}{12} + \dfrac{24}{12}\right]$

$\Sigma x^2 = 770 - [.5+.5+2] = 770-[3] = 767$

2. $\Sigma y^2 = \dfrac{N^3-N}{12} - \Sigma T_x$

$\Sigma y^2 = \dfrac{21^3-21}{12} - \left[\dfrac{2^3-2}{12} + \dfrac{2^3-2}{12} + \dfrac{2^3-2}{12} + \dfrac{2^3-2}{12} + \dfrac{2^3-2}{12} + \dfrac{2^3-2}{12}\right]$

$\Sigma y^2 = 770 - \left[\dfrac{8-2}{12} + \dfrac{8-2}{12} + \dfrac{8-2}{12} + \dfrac{8-2}{12} + \dfrac{8-2}{12} + \dfrac{8-2}{12}\right]$

$\Sigma y^2 = 770 - [.5+.5+.5+.5+.5+.5]$

$\Sigma x^2 = 770 - [3.0] = 767$

3. $r_s = \dfrac{\Sigma x^2 + \Sigma y^2 - \Sigma d^2}{2\sqrt{\Sigma x^2 \Sigma y^2}}$

$r_s = \dfrac{767+767-309}{2\sqrt{(767)(767)}} = \dfrac{1,534-309}{2(767)} = \dfrac{1,225}{1,534} = .799$

This case was a poor selection of data, because the r value was exactly the same using both methods. However, this unusual situation will not always happen, and the procedure used does describe the better method when there are a lot of tied ranks.

3. Uses of Rank Correlation

Rank correlation can be used when the data are not interval but are at least ordinal (rankable). For example, judgement scores are used in some activities, such as manager's ratings, gymnastics, springboard diving, form in a sport, etc.

Although raters may assign points that look like interval or ratio data, the scores are not as accurate as dollars, miles, frequency counts, etc., and the raters may not be using the same starting points or they may be weighing different factors differently, so the distances between their numbers may not be comparable. However, their numbers can be used to obtain a score, and these values can be ranked.

4. Assumptions

The following assumptions are made when using the Spearman-Rho (rank correlation) technique:

1. the numbers are ordinal data or higher.
2. differences between scores can be ranked.
3. gaps between scores are unimportant.
4. the distribution in small samples is discrete and bimodal.
5. position is more important than value of a score.

5. Kuder-Richardson

Reliability coefficients that use some form of correlation coefficient are often used in test construction. Sometimes a teacher wants to know how *reliable or consistent* a certain test is. For example, a written test for a unit in health may be used and updated over a several year period. It would be interesting to know if students could give the same answers to the test each time they took it. That way, a pre - post test might indicate how well the class as a whole mastered subject content, provided the test was valid, reliable and objective.

The usual way to determine reliability or consistency is a test-retest; that is, give the test to a group of students and a short time later administer the test to the same students again. Provided there has been no opportunity to learn test items between the two tests, a comparison of the scores on each test would result in a correlation ratio that could be checked for significance.

The Kuder Richardson reliability coefficient is an internal consistency ratio found with only *one* test administration. It is done by using the following procedure:

1. rank the test scores from high to low.
2. determine the mean and standard deviation.
3. use the Kuder-Richardson formula:

(37)

$$r_{tt} = \frac{N\sigma t^2 - M(N-M)}{\sigma t^2 (N-1)}$$

Where: r_{tt} = *correlation coefficient for total test*

N = *number of items (test questions) on test*

σt^2 = *variance (s^2) of the test*

M = *mean of data*

Using this procedure for finding reliability is not a substitute for the usual test-retest method. But it may be desirable to give an objective written test several times before changing it. Before assigning grades, it might be useful to have some knowledge about the internal test reliability. Giving the test again to the same students may be overkill or it may waste class time. Use r_{tt} to estimate internal reliability and consider the reliability results when assigning grades.

Example: **Table 13.3** displays scores made on a 50 item end of unit exam. What is the reliability ratio?

Table 13.3. Scores Made on a 50 Item Test.

Score (X)	X^2	Score (X)	X^2
47	2,209	35	1,225
47	2,209	34	1,156
46	2,116	33	1,089
43	1,849	33	1,089
41	1,681	33	1,089
41	1,681	33	1,089
41	1,681	33	1,089
40	1,600	32	1,024
40	1,600	32	1,024
39	1,521	31	961
38	1,444	31	961
37	1,369	31	961
36	1,296	28	784
35	1,225	Σ 990	37,022

SOLUTION: $N = 27$; $\overline{X} = 36.67$

1. *Find standard deviation*

   a. $\Sigma x^2 = \Sigma X^2 - \dfrac{(\Sigma X)^2}{N} = 37{,}022 - \dfrac{990^2}{27} = 722$

   b. $s^2 = \dfrac{\Sigma x^2}{N-1} = \dfrac{722}{26} = 27.76$

   c. $s = \sqrt{s^2} = \sqrt{27.76} = 5.27$

2. *Calculate* r_{tt}

$$r_{tt} = \frac{N\sigma t^2 - M(N-M)}{\sigma t^2 (N-1)}$$

$$r_{tt} = \frac{50(5.27)^2 - 36.67(50-36.67)}{(5.27)^2(50-1)}$$

$$r_{tt} = \frac{1{,}388.645 - 488.81}{1{,}360} = .66$$

Entering Table 2 in Appendix B with N-1 df (there is only 1 set of data, not matched pairs, so there are 26 df), an r of .381 is needed for significance at the .05 level (2-tailed test; because 26 df is not listed, n = 25). Since the r found was .66, the data are significantlly different beyond the .05 level, meaning the data are highly correlated, the test results are consistent, and the test is reliable.

The rank difference method discussed in this chapter is a non-parametric technique that makes few assumptions about the data, other than data is rankable. Both parametric and non-parametric methods of correlation are used to determine validity, reliability and objectivity in all kinds of tests - affective, cognitive and physical. Kuder-Richardson is a special case of reliability for written tests.

6. Split-Halves

nother way to determine reliability in a single written test administration is to use a technique called *split-halves*. In the split-halves procedure, a correlation (usually by Pearson) is determined by comparing the number who answered question pairs *correctly*. The following procedure is one way to use this method:

1. use odd questions as X scores, even questions as Y.
2. set up a distribution chart for correlation.
3. determine r the usual way.
4. determine the split-halves r as follows:

$$r_{tt} = \frac{2r_{hh}}{1 + r_{hh}}$$

Where: r_{tt} = *r for total test*

r_{hh} = *r for two halves*

Example: Using the data in Figure 13.1, solve for r_{tt}.

1. $\Sigma x^2 = \Sigma X^2 - \dfrac{(\Sigma X)^2}{N} = 11{,}633 - \dfrac{511^2}{25} = 918.6$

2. $\Sigma y^2 = \Sigma Y^2 - \dfrac{(\Sigma Y)^2}{N} = 9{,}898 - \dfrac{476^2}{25} = 834.96$

3. $\Sigma xy = \Sigma XY - \dfrac{(\Sigma X)(\Sigma Y)}{N} = 10{,}258 - \dfrac{(511)(476)}{25} = 528.56$

4. $r = \dfrac{\Sigma xy}{\sqrt{\Sigma x^2 \Sigma y^2}} = \dfrac{528.56}{\sqrt{(918.16)(834.96)}} = \dfrac{528.56}{875.57} = .60$

5. $r_{tt} = \dfrac{2r_{hh}}{1 + r_{hh}} = \dfrac{2(.60)}{1 + .60} = \dfrac{1.20}{1.60} = .75$

Question #	# Answered Correctly Odd (X)	Even Y	X²	Y²	XY
1 - 2	27	21	729	441	567
3 - 4	9	7	81	49	63
5 - 6	26	19	676	361	494
7 - 8	26	10	676	100	260
9 - 10	17	17	289	289	289
11 - 12	21	18	441	324	378
13 - 14	20	24	400	576	480
15 - 16	25	23	625	529	575
17 - 18	26	26	676	676	676
19 - 20	16	11	256	121	176
21 - 22	11	9	121	81	99
23 - 24	23	23	529	529	529
25 - 26	24	23	576	529	552
27 - 28	25	21	625	441	525
29 - 30	26	26	676	676	676
31 - 32	26	26	676	676	676
33 - 34	25	26	625	676	650
35 - 36	20	10	400	100	200
37 - 38	14	18	196	324	252
39 - 40	23	23	529	529	529
41 - 42	20	17	400	289	340
43 - 44	14	19	196	361	266
45 - 46	18	25	324	625	450
47 - 48	25	20	625	400	500
49 - 50	4	14	16	196	56
Σ	511	476	11,363	9,898	10,258

Figure 13.1. Number Answering Paired Questions Correctly.

r_{tt} is interpreted the same way as described earlier in Chapter 12 on correlation.

Sometimes groups of data need to be compared in order to find out if their means and variances are statistically different. Chapter 14 will analyze data by introducing a "t" test.

APPLICATIONS

Τ he rank difference procedure is used when the data are ordinal. Examples include ratings by managers, coaches, judges, peers, tour groups, tour operators, etc.

When checking a written test for reliability, it may be helpful to use the Kuder Richardson or the split-halves techniques to find a reliability coefficient. These procedures save time over a conventional test-retest method.

Whereas correlation techniques are designed to discover relationships that exist between groups of data, other associations between means and variances may need to be explored. Part 4 examines statistical tools that help to analyze these associations.

Exercises

1. Reproduced below is **Table 12.10**.

Table **12.10**. Hypothetical Millions of Vacation Person-Trips.

Month	1986	1987
1	20.20	20.30
2	21.54	20.76
3	20.50	20.85
4	21.06	21.00
5	20.90	21.18
6	21.80	21.23
7	21.40	21.23
8	21.54	21.27
9	21.80	21.49
10	21.40	21.52
11	21.44	21.62
12	21.40	21.71

a. Change the data to ranks and find and interpret the Spearman Rho correlation coefficient.

b. Draw a scatter plot of the data, using ranks (a line of best fit cannot be drawn).

2. **Table 12.11** is reproduced below.

Table 12.11. Attendance in 1987 and 1988.

Program	1987 (X)	1988 (Y)	Program	1987 (X)	1988 (Y)
1	37	66	12	42	67
2	51	68	13	42	55
3	28	45	14	39	56
4	37	67	15	29	55
5	42	55	16	48	42
6	24	55	17	32	67
7	31	50	18	35	91
8	32	65	19	23	57
9	38	61	20	32	33
10	47	43	21	25	59
11	33	62	22	34	63

a. Change the scores to ranks and find and interpret the Spearman Rho correlation coefficient.

b. Draw a scatter plot of the data using ranks (a line of best fit cannot be drawn).

Shown below is **Table 12.12** reduced to ranks.

Table 12.12. Office Activities.

	Office	Walk-Ins	Mail	Answered	
Week	Local	Tourists	Relocate	Tourists	Other
1	10	5	7	9.5	6
2	12	6	4	3	4
3	7.5	3.5	2	1	3
4	5	3.5	6	4	5
5	1	11	5	7.5	11
6	9	13	11	7.5	7.5
7	14	15	15	15	13
8	7.5	7.5	8	5.5	10
9	3	10	12	9.5	9
10	2	1	1	13	1
11	11	7.5	9	11	7.5
12	15	14	14	5.5	15
13	6	2	13	12	13
14	13	12	10	14	13
15	4	9	3	2	2

3. Using Office Walk-Ins (Local) as X and Mail Answered (Other) as Y in **Table 12.12**, determine and interpret the Spearman Rho correlation coefficient. Draw a scatter plot using the ranks (a line of best fit cannot be drawn).

4. Determine and interpret the Rho correlation coefficient for Walk - Ins (Tourists) and Mail Answered (Tourists) in **Table 12.12**.

Table 13.4. Sociometric Ranks 1 Week Apart.

Subject	Test 1		Test 2	
1	8		19.5	TIE
2	7		11.5	TIE
3	10		3	
4	18.5	TIE	14	
5	16	TIE	19.5	TIE
6	3		9	TIE
7	2		4	
8	5		2	
9	9		6	TIE
10	6		6	TIE
11	18.5	TIE	15	
12	12	TIE	9	TIE
13	16	TIE	9	TIE
14	4		11.5	TIE
15	14		13	
16	12	TIE	19.5	TIE
17	21		16.5	TIE
18	1		1	
19	20		16.5	TIE
20	12	TIE	6	TIE
21	16	TIE	19.5	TIE

The Cowell Personal Distance Scale in **Table 13.4** was designed for high school. In order to check reliability for college use, the ranks were obtained on two separate administrations, 1 week apart.

5.   Using **Table 13.4**, find and interpret the correlation coefficient corrected for ties. Was the test reliable? Explain your answer.

6.   Using the mail answered (other) data in problem 2, and assuming the data are test scores with 100 questions on the test, determine r_{tt} by the Kuder-Richardson Formula. Interpret this statistic. Was the test reliable? Explain your answer.

Reproduced below is **Table 12.14**.

Table 12.14. Chamber of Commerce Walk-Ins and Hotel Occupancy Rate.

	Walk-Ins (X)	Occupancy Rate (Y)
January	37	70%
February	61	65%
March	81	67%
April	97	68%
May	88	67%
June	156	80%
July	203	90%
August	254	92%
September	145	75%
October	131	65%
November	63	60%
December	31	75%

7.   Determine and interpret the Spearman Rho correlation coefficient for Walk-Ins (Tourists) and Mail Answered (Tourists) in **Table 12.14**.

Reproduced below is **Table 12.15**.

Table 12.15. Travel Sales and GNP During the 1980's[1].

Year	Travel Industry Sales	Gross National Product (GNP)
1980	10.8	8.9
1981	11.0	11.7
1982	4.8	3.7
1983	7.9	7.6
1984	7.6	10.8
1985	4.8	6.4
1986	8.8	5.4
1987	10.3	6.7
1988	8.9	7.9
1989	7.2	6.7

[1]U.S. Travel Data Center. (1990). _1990 Outlook for travel and tourism._ Washington: U.S. Travel Data Center.

8.   Determine and interpret the Spearman Rho correlation coefficient for Travel Sales and GNP in **Table 12.15**.

Part 4

Raw Scores

Group Associations

Chapter 14

Measures Involving Variance

1. t Testing

A"t" test is a way to determine if the means of two sets of data are statistically different. To determine this statistic, a difference between the two means is used as the numerator (observed error) and the square root of the sums of each variance, divided by N, is the denominator (expected error). A table (Table 4 in Appendix B) is used to test the significance level of the statistic. Figure 14.1 below illustrates a test that was not significant initially but was significant at a later time. The statistic is found by the following ratio:

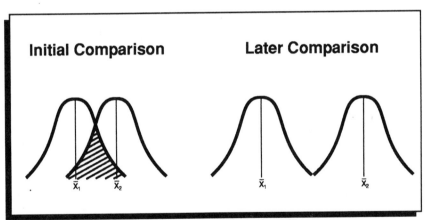

Figure 14.1. A Graphic Representation of a t Test.

$$(38) \qquad t = \frac{|\overline{X}_1 - \overline{X}_2|}{\sqrt{\dfrac{s_1^2}{N_1} + \dfrac{s_2^2}{N_2}}}$$

Example: **Table 14.1** lists data from 2 randomly assigned groups:

Table 14.1. Number of Visits to a Community Pool.

Before Learning to Swim		After Learning to Swim	
X_1	X_1^2	X_2	X_2^2
2	4	4	16
4	16	5	25
5	25	8	64
7	49	8	64
9	81	7	49
8	64	9	81
3	9	1	1
Σ 38	248	42	300

Solution, using Equation 38:

1. $\Sigma x_1^2 = \Sigma X_1^2 - \dfrac{(\Sigma X_1)^2}{N_1} = 248 - \dfrac{38^2}{7} = 41.72$

2. $s_1^2 = \dfrac{\Sigma x_1^2}{N_1 - 1} = \dfrac{41.72}{7-1} = 6.95$

3. $\Sigma x_2^2 = \Sigma X_2^2 - \dfrac{(\Sigma X_2)^2}{N_2} = 300 - \dfrac{42^2}{7} = 48$

4. $s_2^2 = \dfrac{\Sigma x_2^2}{N_2 - 1} = \dfrac{48}{7-1} = \dfrac{48}{6} = 8.00$

5. $\bar{X}_1 = \dfrac{38}{7} = 5.43; \quad \bar{X}_2 = \dfrac{42}{7} = 6.00$

(continued on next page)

6. $t = \dfrac{|\bar{X}_1 - \bar{X}_2|}{\sqrt{\dfrac{s_1^2}{N_1} + \dfrac{s_2^2}{N_2}}} = \dfrac{|5.43 - 6.00|}{\sqrt{\dfrac{6.95}{7} + \dfrac{8.00}{7}}} = \dfrac{0.57}{\sqrt{0.99 + 1.14}}$

$t = \dfrac{0.57}{\sqrt{2.13}} = \dfrac{0.57}{1.46} = .390$

To explain the statistic, the df (n) must be found. Unlike r, n for t = N-1 for a matched pair of data. Since there are 7 scores, n = 7-1=6.

Entering **Table 4** in Appendix B with 6 n (df), go across row 6 to the column needed for significance (assume p = .05). For a one-tail test, find the column labeled .1 (.05 doubled), and read the value needed directly from the table (1.943). Since the value of t was only .390, and the table value needed for significance at the .05 level was 1.943, the means are not significantly different.

If a two-sided test is desired, read directly from the table (for example, for a two-tailed test at the .05 level, go to the .05 column and read the value listed. With 6df or 6n, a value of 2.447 is needed).

2. t With Repeated Measures

hen using the *same* subjects on a repeated measure, such as a pre-post test, there is a high chance of correlated data. A person who is used twice on the same measure has scores that are related, and the probability of at least some correlation is apparent. For that reason, it may be helpful to compute r, and correct for it in the "t" statistic. The correction factor is minus $2rs_{\bar{x}_1}s_{\bar{x}_2}$, and is under the denominator radical (see equation 39 on the next page).

Example: Visits by 10 clients to a community center were recorded. After a one week media campaign, they were recorded again. Was there any statistical difference in group scores from pre to post campaign? **Table 14.2** summarizes the data.

$$(39) \qquad t_{corr} = \frac{|\bar{X}_1 - \bar{X}_2|}{\sqrt{\dfrac{s_1^2}{N_1} + \dfrac{s_2^2}{N_2} - 2r\,(s_{\bar{x}_1})(s_{\bar{x}_2})}}$$

Table 14.2. Number of Visits made to a Community Center.

Subject	X_1(Pre)	X_2 (Y)	X_1^2	X_2^2	XY
1	1	3	1	9	3
2	2	4	4	16	8
3	3	5	9	25	15
4	4	2	16	4	8
5	5	2	25	4	10
6	2	4	4	16	8
7	3	4	9	16	12
8	3	7	9	49	21
9	4	7	16	49	28
10	6	8	36	64	48
Σ	33	46	129	252	161

Solution using Equation **38** *for t (not t_{corr}):* $(\bar{X}_1 = 3.3;\ \bar{X}_2 = 4.6)$

1. $\Sigma x_1^2 = \Sigma X_1^2 - \dfrac{(\Sigma X_1)^2}{N_1} = 129 - \dfrac{33^2}{10} = 129 - 108.9 = 20.1$

2. $s^2 = \dfrac{\Sigma x_1^2}{N-1} = \dfrac{20.1}{10-1} = \dfrac{20.1}{9} = 2.23$

3. $s_1 = \sqrt{s^2} = \sqrt{2.23} = 1.49$

4. $s_{\bar{x}_1} = \dfrac{s_1}{\sqrt{N_1}} = \dfrac{1.49}{\sqrt{10}} = \dfrac{1.49}{3.16} = .47$

5. $\Sigma x_2^2 = \Sigma X_2^2 - \dfrac{(\Sigma X_2)^2}{N_2} = 252 - \dfrac{46^2}{10} = 252 - \dfrac{2,116}{10} = 252 - 211.6$

$\Sigma x_2^2 = 40.4$

6. $s_2^2 = \dfrac{\Sigma x_2^2}{N_2} = \dfrac{40.4}{10-1} = \dfrac{40.4}{9} = 4.49$

7. $s_2 = \sqrt{s^2} = \sqrt{4.49} = 2.12$

8. $s_{\bar{x}_2} = \dfrac{s_2}{\sqrt{N_2}} = \dfrac{2.12}{\sqrt{10}} = \dfrac{2.12}{3.16} = .67$

9. $t = \dfrac{|\bar{X}_1 - \bar{X}_2|}{\sqrt{\dfrac{s_1^2}{N_1} + \dfrac{s_2^2}{N_2}}} = \dfrac{|3.3 - 4.6|}{\sqrt{\dfrac{2.23}{10} + \dfrac{4.49}{10}}} = \dfrac{1.3}{\sqrt{\dfrac{6.72}{10}}} = \dfrac{1.3}{\sqrt{.67}} = 1.59$

Using a one-tail test for matched pairs (dependent data) with N-1 (9) df, the table value for the .05 level is found to be 1.833. Since the t obtained is 1.59, the means *are not significantly different* at the .05 level. However, when correcting for correlation, the value becomes:

1. $\Sigma xy = \Sigma XY - \dfrac{(\Sigma X)(\Sigma Y)}{N} = 161 - \dfrac{(33)(46)}{10} = 161 - \dfrac{1,518}{10}$

$\Sigma xy = 161 - 151.8 = 9.2$

THEN, using Equation 39 for t_{corr} (see next page)

2. $r = \dfrac{\Sigma xy}{\sqrt{\Sigma x^2 \Sigma y^2}} = \dfrac{9.2}{\sqrt{(20.1)(40.4)}} = \dfrac{9.2}{\sqrt{812.04}} = \dfrac{9.2}{28.50} = .36$

3. $t_{corr} = \dfrac{|3.3-4.6|}{\sqrt{\dfrac{2.23}{10} + \dfrac{4.49}{10} - 2(.36)(.47)(.67)}} = \dfrac{1.3}{\sqrt{(.67)-(.23)}} = \dfrac{1.3}{\sqrt{.44}}$

$t_{corr} = \dfrac{1.3}{.66} = 1.97$

The table value for a one-tail test at the .05 level remains at 1.833, but the new t value is now 1.97, showing the means, *when corrected for correlation, are significantly different* at the .05 level of confidence.

When using t corrected for correlation, there is an assumption that N is the same for both sets of data. Therefore, the procedure will not work when the number of scores for both groups are different.

Discussion. In the problem completed, the hypothesis test is trying to determine if mean 1 = mean 2. A one-tail test can be made because if clients are able to record a certain number of visits, then are given a media campaign (most likely including new information) it is expected they will want to visit the center more. Also, the positive atmosphere generated by the media campaign in general should result in more motivation by the client and better attention by center personnel, increasing chances for more visits after the campaign.

Since it is expected the scores will be better the second week, a one-tail test can be made on the positive side of the curve. In any case, it is necessary to consider which direction, if any, the scores should take. If there is sufficient reason to predict a positive or negative direction, a one-tail test can be made. If it is not known what direction the scores should take, a two-sided test is given.

So far, the test of significance has focused on two means that were dependent on each other; that is, the scores were matched sets of data made by the same client on two comparisons. If the data are *independent* - that is, one score does not influence another score (not comparing the same clients on different occasions), the t statistic is determined as before, but the degrees of freedom change. For example, a design may be used where data made by one group are compared with data made by a different group, where no clients attend both groups. In this case the data are independent, and the t statistic is found by using Equation 38, not equation 39.

Assuming the data in the example given for visiting the community center were not recorded by the same clients, but was given to two different groups as a post test in which two different media campaigns were employed, and assuming further that none of the same clients were in both groups, the t statistic remains at 1.59, but df now become N -1 for each set of data, or 20 -2 for the total. Entering the table with 18 df, the t necessary for significance at the .05 level is now 1.734. Since the t obtained is 1.59, the data are not different at the .05 level. Since the data are independent, t_{corr} is not an appropriate procedure for comparison in independent groups..

3. Assumptions of t

In using t as an *independent* measure, the following assumptions are made about the data:

1. Independence. No score should influence any other score.
2. Random sample.
3. Normal distribution. If both distributions *approximate* normality, use t with caution.
4. Homogenous variance. May be violated if $N_1 = N_2$.
5. Interval data or higher.

The t test is not often used today, because a more powerful statistic called analysis of variance (ANOVA), that has most of the same assumptions, can be used instead. Chapter 15 will discuss ANOVA.

APPLICATIONS

A t test can be used to compare group means if the data assumptions are met. If the data are not *independent*, such as when comparing pre-post tests using the same subjects, use t $_{corr}$.

EXERCISES

1. A media campaign posted the following pre and post scores:

Table 14.3. Average Number of Daily Visits.

Client	Pre (X_1)	Post (X_2)
1	1.20	1.36
2	1.27	1.72
3	1.05	.90
4	.77	.95
5	.64	1.18
6	.75	1.18
7	.95	1.09
8	.72	.61
9	.97	1.18
10	1.27	1.68
11	.86	1.36
12	.82	1.09
13	1.75	2.00

Using t_{corr}, was there a difference in pre to post attendance? (Find $\bar{X}_1$, $\bar{X}_2$, ΣX_1, ΣX_2, ΣX_1^2, ΣX_2^2, ΣXY, Σx^2, Σy^2, Σxy, r, s_1^2, s_2^2, s_1, s_2, $s_{\bar{x}_1}$, $s_{\bar{x}_2}$, t_{corr}). Interpret this t. What do the data mean?

2.   Ten clients from two different fitness swim groups were randomly chosen to participate in an experiment. They were tested initially on the number of lengths of the pool they could swim in twelve minutes, and no statistical differences were found between the two means. The subjects were ranked and matched, based on their pre-test scores and were then exposed to two different conditioning methods. **Table 14.4** shows the post test results.

Table 14.4. Number of Lengths Swam in 12 Minutes.

Matched Pair	Method # 1 (X_1)	Method # 2 (X_2)
1	12	16
2	15	22
3	7	8
4	3	12
5	20	37
6	9	12
7	2	7
8	5	7
9	9	6
10	1	1

Using **Table 14.4**, was there a difference in the performance of the two groups? Employ a t test to determine significance levels.

3. Twenty poll takers were randomally assigned to two groups prior to the start of a data collection experiment. The scores of the two groups were compared on the number of clients interviewed in one day to see if they differed statistically. **Table 14.5** lists the number of clients interviewed.

Table 14.5. Number of Clients Interviewed in One Day.

Control Group	Experimental Group
157	268
92	173
243	187
160	67
181	329
81	264
200	232
225	137
192	249
220	148

Using **Table 14.5**, was there a difference between the control and experimental groups before the experiment began? Explain.

4.   The following data was obtained from a program comparison:

Table 14.6.   Number of Registered Participants.

Program	1994	1995	Program	1994	1995
1	12	14	18	19	20
2	15	18	19	18	19
3	16	22	20	19	24
4	15	22	21	12	21
5	16	19	22	10	21
6	14	21	23	13	17
7	11	18	24	19	23
8	13	25	25	17	17
9	11	15	26	17	18
10	12	29	27	15	19
11	10	17	28	14	19
12	20	19	29	13	22
13	14	19	30	9	16
14	15	17	31	12	20
15	14	19	32	14	19
16	19	20	33	14	17
17	18	19	34	10	20

Using Table 14.6, find t_{corr} and analyze the significance level (find $\bar{X}_1$, $\bar{X}_2$, ΣX_1, ΣX_2, ΣX_1^2, ΣX_2^2, ΣXY, Σx^2, Σy^2, Σxy, r, s_1^2, s_2^2, s_1, s_2, $s_{\bar{x}_1}$, $s_{\bar{x}_2}$, t_{corr}).

A hotel was dissatisfied with their percent of occupancy in 1994. A marketing plan was started early in 1995 and the data were compared with 1994 to see if there were any statistical differences. **Table 14.7** lists the data.

Table 14.7. Occupnacy Rates in 1994 and 1995.

Month	Rate (1994)	Rate (1995)
Jan	70%	65%
Feb	65%	70%
Mar	67%	74%
Apr	68%	72%
May	67%	83%
Jun	80%	85%
Jul	90%	90%
Aug	92%	95%
Sep	75%	83%
Oct	65%	69%
Nov	60%	67%
Dec	75%	83%

5.   Determine the following statistics:

   a.   Mean 1994 & 1995.
   b. Σx^2 for 1994 and 1995.
   c.   Variance for 1994 and 1995.
   d.   t.
   e.   Interpret t.

10 state parks were selected for a different referral system by the state tourism department. The 10 parks involved had intensive follow-up to tourism inquires using a special 800 number. Head counts of visitors were kept during 1994 and 1995. The data are listed in **Table 14.8** below.

Table 14.8. Number of Visitors (Millions) in 10 Parks.

Park	1994 (in Millions)	1995 (Millions)
A	9	21
B	10	19
C	18	20
D	14	13
E	18	15
F	5	20
G	8	22
H	11	25
I	12	17
J	13	10

6.   Determine the following statistics:

    a.   Mean 1994 & 1995.
    b. Σx^2 for 1994 and 1995.
    c.   Variance for 1994 and 1995.
    d.   t.
    e.   Interpret t.

Chapter 15

ANOVA

1. One Way Analysis of Variance

I n using a t test, the difference between means was analyzed. Of course, variance is a part of the t procedure, so it should come as no surprise that there must be some way to look at the variances statistically. The F test is a way to analyze the variances, and it is interesting to note that when there are only two sets of data, $t^2 = F$.

For a one-way analysis of variance the following assumptions are considered:

1. normality.
2. equal variance.
3. independence.
4. random sample.
5. equal means.
6. interval or ratio data.
7. equal r for correlated observations (same subjects X different treatments).

The F test is a test of *variance* between a score and it's mean. Subjects are exposed to different treatments, and if there is no subject difference, then error is assumed as due to the treatment. By exposing subjects to different treatments, the overall variance increases, but individual variance remains the same. It is the position of the means that changes, not the overall variance *within* the group. However, the overall variance *between* groups does change

Example. A problem from Chapter 14 that violates several ANOVA assumptions but is used because of it's simplicity is restated below:

The attendance of 10 clients visiting a community center were recorded. After a one week marketing campaign, they were recorded again. Was there any statistical difference in group scores from pre to post test? Data from **Table 14.1** is reproduced below.

Table 14.1. Number of Visits to a Community Center.

Client	Pre (A_1)		Post (A_2)		X_T	X_T^2
	X_1	X_1^2	X_2	X_2^2		
1	1	1	3	9		
2	2	4	4	16		
3	3	9	5	25		
4	4	16	2	4		
5	5	25	2	4		
6	2	4	4	16		
7	3	9	4	16		
8	3	9	7	49		
9	4	16	7	49		
10	6	36	8	64		
Σ	33	129	46	252	79	381

To solve for ANOVA, different Σx^2 need to be found. Many texts refer to Σx^2 with the symbol SS. This text will use the symbol Σx^2 in this chapter so that a link with procedures in prior chapters will be more readily understood. However, the source tables in this chapter and discussions in Chapter 16 will use the SS symbol to refer to the sum of deviations squared.

In the present problem, sums of deviations squared total (Σx_T^2), level 1 (Σx_1^2) (column 1), level 2 (Σx_2^2) (column 2), expected (also called error or within) $(\Sigma x^2 E)$ and treatments $(\Sigma x^2 TR)$ are found in order to compute a statistic called the F ratio. The F ratio is then tested for significance in a special table (Appendix B, *Table 5*) in ways similar to r and t. Equation 40 describes the procedure used.

$$(40) \quad F = \frac{\dfrac{\Sigma x^2 TR}{df_{TR}}}{\dfrac{\Sigma x^2 E}{df_E}} = \frac{Between\ mean\ square}{Within\ mean\ square}$$

Using the data in **Table 14.1**, the following procedure is used to solve Equation 40.

1. $\Sigma x^2{}_T = \Sigma X^2{}_T - \dfrac{(\Sigma X_T)^2}{N_T} = 381 - \dfrac{79^2}{20} = 381 - 312.05 = 68.9$

2. $\Sigma x_1^2 = \Sigma X_1^2 - \dfrac{(\Sigma X_1)^2}{N_1} = 129 - \dfrac{33^2}{10} = 129 - 108.9 = 20.1$

3. $\Sigma x_2^2 = \Sigma X_2^2 - \dfrac{(\Sigma X_2)^2}{N_2} = 252 - \dfrac{46^2}{10} = 252 - \dfrac{2,116}{10}$

   $\Sigma x_2^2 = 252 - 211.6 = 40.4$

4. $\Sigma x^2 E = \Sigma x_1^2 + \Sigma x_2^2 = 20.1 + 40.4 = 60.5$

5. $\Sigma x^2 TR = \Sigma x^2{}_T - \Sigma x^2 E = 68.95 - 60.5 = 8.45$

6. $F = \dfrac{\dfrac{\Sigma x^2 TR}{df_{TR}}}{\dfrac{\Sigma x^2 E}{df_E}} = \dfrac{\dfrac{8.45}{1}}{\dfrac{60.5}{18}} = \dfrac{8.45}{3.36} = 2.51$

Procedure 6 above may also be found by drawing a source table, as in **Table 15.1** on the next page.

Table 15.1. Source Table for Number of Clients Attending Two Centers.

Source	df	SS (Σx^2)	MS (Mean Square)	F	p
Total	19	68.95			
TR (Between)	1	8.45	8.45	2.51	n.s.
E (Within)	18	60.50	3.36		

In order to determine whether or not the F is significant, it is necessary to use a table of F (**Appendix B, Table 5**). Entering the table across the top with the greater mean square (TR = 8.45; df = 1), find the column labeled 1 df. Then go down the column vertically with the error df (18). The .05 level is the table value listed at the top, while the .01 level is the table value listed at the bottom. Since the table value needed for significance at the .05 level is 4.41, the F obtained of 2.51 is not significant (n.s.).

Earlier, it was stated that t^2 = F. In the example above, t was found to be 1.59 (page 223). 1.59^2 = 2.53. The .02 difference is due to rounding errors.

Degrees of Freedom (df). Determining the degrees of freedom for the analysis of variance requires some thought. The degrees of freedom is always N-1, but determining what the value for N should be is confusing.

For df total, recall that there are two sets of data. The data includes 20 **total** scores. In that group of 20 scores, all but one score could be any value at all. However, the last score must be some value that will total out to be the final value of all the scores listed, so one score cannot vary. Therefore, total df = 20-1 = 19.

For treatments, there were two sets of data or two treatments - the initial set of scores, plus the second set. Therefore, df for treatments is 2-1 or 1. Another way of looking at it is to realize that an initial set of scores was obtained, and something different was done (one treatment) before the next set of scores were obtained. But the easiest way is to count the columns of data and subtract 1 for treatments df. Treatment data is known as *between* groups.

The error, or *within* df, has to be whatever is left. Since there were 19 total df, and one treatment df, the df expected (E) has to be 19-1 or 18. The total must equal the beginning df, which was 20-1 or 19.

Summary of Steps. It is usually much easier to draw a source table when determining F in order to find the proper degrees of freedom associated with each sum of squares. The source table makes the data easier to use, because after the df total is found, treatments df can be subtracted from df total to find df expected. To summarize, in order to find F in one-way analysis of variance:

$$1.\ \Sigma x^2{}_T = \Sigma X^2{}_T - \frac{(\Sigma X_T)^2}{N_T}$$

$$2.\ \Sigma x_1^2 = \Sigma X_1^2 - \frac{(\Sigma X_1)^2}{N_1}$$

$$3.\ \Sigma x_2^2 = \Sigma X_2^2 - \frac{(\Sigma X_2)^2}{N_2}$$

$$4.\ \Sigma x^2 E = \Sigma x_1^2 + \Sigma x_2^2 ... + \sum_{k=1}^{j}$$

$$5.\ \Sigma x^2 TR = \Sigma x^2{}_T - \Sigma x^2 E$$

$$6.\ F = \frac{\dfrac{\Sigma x^2 TR}{df}}{\dfrac{\Sigma x^2 E}{df}}$$

It should be pointed out that there are other ways of computing the sum of squares for treatments. Since the method cited is easy to understand and ties in nicely with the concept of sum of deviations squared, it will be used consistently throughout the remainder of the text.

More Than Two Treatments. A one-way analysis of variance can also be done with more than two treatments. That is why the procedure in step 4 on the previous page indicated *summing the Σx^2 from 1 to j* (j = number of treatments). The procedure followed is the same as with two data sets. The only difference is in finding $\Sigma x^2 E$. To find the $\Sigma x^2 E$ with more than two data sets, add the Σx^2 for each individual set of scores. For example, if there were four data sets, add the total of the Σx^2 for all four sets of data together to find $\Sigma x^2 E$.

Example: clients in a fitness class were randomly assigned to three different groups (A_1, A_2, and A_3). Each group was tested on the number of sit-ups they could do in 10 seconds, and no difference between groups was found. Each group then received different instruction designed to improve abdominal strength. At the end of six weeks, their sit-up scores were listed in **Table 15.2**.

Table 15.2. One-Way, Three Group Design.

Level A_1		Level A_2		Level A_3			
X_1	X_1^2	X_2	X_2^2	X_3	X_3^2	X_T	X_T^2
4	16	5	25	9	81		
3	9	4	16	2	4		
2	4	1	1	6	36		
2	4	3	9	8	64		
5	25	7	49	9	81		
8	64	7	49	10	100		
7	49	4	16	8	64		
7	49	8	64	12	144		
9	81	10	100	13	169		
2	4	4	16	12	144		
Σ 49	305	53	345	89	887	191	1537

Was there a difference in the methods of instruction?

Solution:

1. $\Sigma x^2{}_T = 1{,}537 - \dfrac{191^2}{30} = 1{,}537 - 1{,}216.03 = \mathbf{320.97}$

2. $\Sigma x_1^2 = 305 - \dfrac{49^2}{10} = 305 - 2401. = \mathbf{64.9}$

3. $\Sigma x_2^2 = 345 - \dfrac{53^2}{10} = 345 - 280.9 = \mathbf{64.1}$

4. $\Sigma x_3^2 = 887 - \dfrac{89^2}{10} = 887 - 792.1 = \mathbf{94.9}$

5. $\Sigma x^2 E = (\Sigma x_1^2 + \Sigma x_2^2 + \Sigma x_3^2) = \qquad \mathbf{223.9} \qquad \mathbf{-223.90}$

6. $\Sigma x^2 TR = (\Sigma x^2{}_T - \Sigma x^2 E) = \qquad\qquad\qquad \mathbf{97.07}$

Construct a source table.

Table 15.3. Source Table for One-Way, Three Group Design.

Source	df	SS	MS	F	p
Total	29	320.97			
TR	2	97.07	48.54	5.86	.01
E	27	223.90	8.29		

Discussion. With 2 and 27 degrees of freedom, the data are significant at the .01 level. However, the analysis of variance is gross. There is no way to know which levels (sets of data) are the significant ones. By inspection, it looks like the data in A_3 is considerably different, but it is possible that A_1 & A_2, A_1 & A_3, A_2 & A_3, or A_1, A_2 or A_3 alone were statistically different. If the statistic is to be meaningful, further analysis is needed.

Duncan's Multiple Range Analysis. One way to address the problem is to use Duncan's Multiple Range Analysis. This method uses means to find out which mean, if any, is statistically different from any other. Some additional information is required. The mean of each data set is needed, as is the $s_{\bar{x}}$ for error variance. Then a chart of the means are compared with a special table value called the *shortest studentized range*. This range can then be tested for significance.

1. $\bar{X}_1 = \dfrac{49}{10} = 4.9; \ \bar{X}_2 = \dfrac{53}{10} = 5.3; \ \bar{X}_3 = \dfrac{89}{10} = 8.9$

2. $s_{\bar{x}}E = \dfrac{s}{\sqrt{N}} \ ; \ s = \sqrt{MS \ (within)} \ or \ MS \ (expected) = \sqrt{8.29} = 2.88$

   $N = Number \ of \ observations \ on \ which \ mean \ is \ based$

   $s_{\bar{x}}E = \dfrac{2.88}{\sqrt{10}} = .91$

3. *Construct a table of means like Table* **15.4**

Table 15.4. Table of Means Using Three Levels.

	$\bar{X}_1$ 4.9	$\bar{X}_2$ 5.3	$\bar{X}_3$ 8.9	Shortest Studentized Range	Table Value
$\bar{X}_1$ 4.9	-	0.4	4.0	2.66	R_2 2.92
$\bar{X}_2$ 5.3		-	3.6	2.79	R_3 3.07
$\bar{X}_3$ 8.9			-	-	-
	1	2	3		
1		X			
2					

In **Table 15.4**, it should be obvious that with three means, only two need to be listed in the rows. In all cases, one less than the number of means is usually listed. The additional row ($\bar{x}_3$) is shown only for purposes of explanation. The means **must** be listed from low to high, right to left horizontally, and low to high, descending vertically as shown in **Table 15.4**.

Using Appendix B, **Table 6B**, the column labeled table value is entered with k means and df. For the .05 level, choose the table labeled α=.05. Go down the column labeled df with the df expected (E) (df = 30-3 = 27. There are no 27 df in the table. However, 30 and 24 df are listed. By inspection, 30 df requires a *lower* table value than 24 df. Use the more rigorous test, which by inspection is 24 df.).

Enter the k column with 2 and copy the R_2 table value, 2.92 on the chart. In the same way, the next chart entry is labeled R_3 with a value of 3.07. Continue horizontally across the chart until all table values for the k means are determined. The chart will always start with R_2 as the first table value.

To find the Shortest Studentized Range, multiply the *table values* by $s_{\bar{x}}E$ found earlier (.91). Now the significance test can be obtained.

In interpreting the chart, enter with the Shortest Studentized Range. To be significantly different, any mean difference listed on the row has to equal or exceed exceed the studentized range for the number of means used to compare that row. On the row labeled $\bar{x}_1$, three means are used for comparison purposes ($\bar{x}_1$, *with* $\bar{x}_2$ *and* $\bar{x}_3$). Therefore, the R_3 table value and corresponding shortest studentized range (2.79) is used. Any mean difference that equals or exceeds 2.79 on row $\bar{x}_1$ is significantly different. Inspection reveals $\bar{x}_3$ (difference 4.0) is different. Continuing left, $\bar{x}_2$ (difference = 0.4) is not significant. Thus an X is entered in the chart where mean 2 intersects with mean 1 to indicate the difference is not significant.

The next comparison is of 2 means, $\bar{x}_2$ *with* $\bar{x}_3$. Since only 2 means are involved, the R_2 values are used. This comparison must equal or exceed a studentized range of 2.66. Hence, the difference, 3.6 is significant, and no X is entered in the mean 2 row.

Discussion. The data indicate the variance is different at the .01 level, and further analysis shows $\bar{x}_3$ is different when compared to both $\bar{x}_1$ and $\bar{x}_2$. Obviously, the method used in $\bar{x}_3$ produced more sit-ups than the other methods, provided all other factors were controlled.

Scheffe' Method. Another way to test the significance of means is by a technique defined by Sheffe'. Using this method requires testing ANOVA at the desired alpha level, then comparing the means using the following technique:

$$(41) \qquad I = \left[\sqrt{(K-1)(F_\alpha)}\right]\left[\sqrt{MS_W\left(\frac{1}{n_1}+\frac{1}{n_2}\right)}\right]$$

Where

I = difference between means (.05. .01 F test).

K = number of groups.

F_α = F value for α(alpha) level desired.

MS_W = mean square within (ANOVA Σx^2 TR).

n_1 = number of scores in group 1.

n_2 = number of scores in group 2. Different n's are o.k..

Using data obtained from **Table 15.2** in equation 41:

$$I_{.01} = \left[\sqrt{(3-1)(5.49)}\right]\left[\sqrt{(8.29)\left(\frac{1}{10}+\frac{1}{10}\right)}\right]$$
$$I_{.01} = \left[\sqrt{(2)(5.49)}\right]\left[\sqrt{1.66}\right]$$
$$I_{.01} = [3.31]\ [1.29]$$
$$I_{.01} = 4.27.$$

What the data reveal is that any two means that differ by more than 4.27 are significantly different. Notice from **Table 15.4**, no means were statistically different with the Scheffe' method, even though F was significant. Duncan's analysis is the more liberal of these two methods.

2. Randomized Group Design

The randomized group design is analyzed the same way as one-way ANOVA, but the *organization of the data is random in each row*. In the model discussed in **Table 15.1**, ten clients were assigned randomally to each treatment, but the clients were exposed to each treatment on the same row. This method violates the assumption of independence, but the randomized group design eliminates this bias.

In the randomized group design, clients are assigned randomally to each treatment, as before. However, clients are now assigned randomally to each **row**, first at one level (A_1), then another (A_2). In this way, independence is gained.

As an example, the problem in **Table 14.1** (page 243) lists 10 clients. In the randomized group design, subjects would randomly be assigned to each treatment, and the column heading "subject" would be replaced by a new heading called "observations." At level one (A_1), each client would be assigned randomly to any of the 10 rows. At level two, the same random assignment would occur, *rather than the same client repeating level two in the same row*. Analysis is carried out the same way as in simple one-way ANOVA for any number of columns or treatments.

3. Randomized Block or Repeated Measures Design

A randomized block design is analyzed in much the same way as one-way ANOVA. One difference is in the way the experiment is designed.

A first consideration is in how many levels of A or treatments are to be included. In the case of the problem in **Table**

15.2, there were three levels. *Blocks or rows of level B* are assigned, based on the number of treatments. If there are three treatments, three blocks (rows) are needed. The number of subjects required are equal to the number of blocks X treatments. For example, if there will be 3 blocks or rows, 9 clients (3 rows by 3 columns or treatments) are needed.

The 9 clients are tested and ranked on some variable, high to low, such as sit-ups. Then the first three clients are assigned to block 1, the second three to block two, etc. Within the block or row, the three clients are randomly assigned to each treatment. Analysis of variance is carried out in the same manner as one-way ANOVA, except that Σx^2 for blocks is found, and the sum of squares for determining F is blocks X treatments, instead of error.

$\Sigma x^2 T$, E, and TR are found in the usual way. Then each row SS (Σx^2) are found by:

$$\Sigma x^2{}_{row} = \Sigma X^2{}_{row} - \frac{(\Sigma X_{row})^2}{N_{row}}$$

Next, the SS for each row is added and subtracted from $\Sigma x^2 T$ to obtain the $\Sigma x^2{}_{block\ X\ treatments}$. The $\Sigma x^2{}_{block\ X\ treatments}$ are used in place of the error sum of squares. A source table is drawn and the rest of the analysis is completed in the usual way.

Example: using number of clients recruited as a criterion, three new identical brochures were commercially designed; control (A_1) that used no color, two colors (A_2), and 3 coliors (A_3). Since there are three levels, three blocks were needed. Then nine tour guides were randomly selected from a large tour operator list. They were ranked on the number of new clients recruited in one week, using a brochure designed in-house (see **Table 15.5**)

Table 15.5. Pre-Test Number of New Clients Recruited in One Week.

Subject	1	2	3	4	5	6	7	8	9
Score	6	5	5	4	2	1	1	0	0

Next, subjects 1, 2 and 3 were assigned to block 1; 4, 5 and 6, block 2; and 7, 8 and 9, block 3. Using a table of random numbers, each subject was assigned to a different treatment level within each block (see **Table 15.6** below).

Table 15.6. Subjects Randomized in Three Treatments.

Subject Number			
Block	A_1	A_2	A_3
1	2	3	1
2	6	5	4
3	9	8	7

Recall that randomization is done using drawing with replacement. Thus, if a number is drawn more than once, it is replaced in the group and the drawing continues until a different number is chosen. Therefore, the cells were completed with repeated drawings.

Finally, tour guides were given the new brochures and measured on the criterion. Results of the experiment are shown in **Table 15.7**.

Table 15.7. Number of Clients Recruited Using New Brochures.

Block	A_1		A_2		A_3			
	X_1	X_1^2	X_2	X_2^2	X_3	X_3^2	X_T	X_T^2
1	8	64	8	64	9	81		
2	4	16	5	25	7	49		
3	3	9	4	16	0	0		
Σ	15	89	17	105	16	130	48	324

Procedure: It will be necessary to find sums of deviations squared for total, X_1, X_2, X_3, expected (within), treatments (between), blocks, and blocks times treatments.

1. $\Sigma x^2_T = 324 - \dfrac{48^2}{9} = 68$

2. $\Sigma x_1^2 = 89 - \dfrac{15^2}{3} = 14$

   $\Sigma x_2^2 = 90 - \dfrac{16^2}{3} = 8.67$

   $\Sigma x_3^2 = 130 - \dfrac{16^2}{3} = 44.67$

3. $\Sigma x^2 E = \Sigma x_1^2 + \Sigma x_2^2 + \Sigma x_3^2 = 67.34$

4. $\Sigma x^2 TR = \Sigma x^2_T - \Sigma x^2 E = 68 - 67.34 = .66$

5. $\Sigma x^2_{blocks} = \Sigma X^2_T \ (block\ 1) + \Sigma X^2_T \ (block\ 2)$

   $+\Sigma X^2_T \ (block\ 3) - \dfrac{\Sigma X^2_T}{N_T} = \dfrac{25^2}{3} + \dfrac{16^2}{3} + \dfrac{7^2}{3} - \dfrac{48^2}{9}$

   $= [208.33 + 85.33 + 16 - 53] - 256$

   $\Sigma x^2_{blocks} = 309.99 - 256 = 53.99$

6. $\Sigma x^2_{blocks\ X\ treatments\ (B\ X\ T)} = \Sigma x^2_T - [\Sigma x^2_{blocks} + \Sigma x^2 TR]$

   $\Sigma x^2_{B\ X\ T} = 68 - [53.99 + .66] = 68 - 54.65 = 13.35$

Table 15.8. Source Table Using New Brochures.

Source	df	SS	MS	F	p
Total	8	68			
TR	2	0.66	0.34	0.10	n.s.
Blocks	2	53.99	26.95		
B X T	4	13.35	3.34		

Discussion. As in the cases cited earlier, treatments refers to the *between* variance, and in this case, block X treatments is used as the *within* variance. Notice the $\Sigma x^2 E$ is not used in the source table. Instead, $\Sigma x^2_{\text{blocks X treatments}}$ is used to find F (ie., F = $MS_{\text{treatments}}$ divided by $MS_{\text{blocks X treatments}}$). For repeated measures, *subjects* replace blocks and F = $MS_{\text{treatments}}$ divided by $MS_{\text{subjects X treatments}}$.

Degrees of freedom are found in the usual way. Total df is N-1 = 9-1 = 8. Since there are three treatments, df for treatments = 3-1 = 2. There are three blocks, so df for blocks is 3-1 = 2. Degrees of freedom for blocks X treatments must be the remainder, or 4.

Any time the F ratio is less than one, there is no need to look up p in the table. However, if the ratio had been >1, determining p from a table of F is the same as for one-way ANOVA.

In comparing the randomized group design with the randomized block design, the within SS is error for the group design, while the block X treatments SS is the error for the block design.

APPLICATIONS

It is important to consider data characteristics as mentioned at the beginning of this chapter. In assessing the data, design the experiment before data collection begins. If groups are going to be compared, check to see the groups are statistically equivalent before starting the experiment. If the means are to be analyzed, be sure the means are initially the same. A "t" test may be used for this purpose.

If the experiment is using variance measures be sure the initial variance is the same. A statistic called Hartley's test or Cochran's test may be used. These tests are simple to use but require special tables for analysis[1].

Chapter 16 will show how to use ANOVA when more than one level are analyzed.

[1] Meyers, J.L. (1969). Fundamentals of experimental design (p. 73). Boston: Allyn and Bacon.

EXERCISES

Fifteen programs organized by two different supervisors (N = 30) were randomly selected to participate in a study to determine if there were any demographic differences in client attendance over a six month time period. Except for the clients, the known conditions were the same; so far as could be determined, only the supervisiors of these programs were different. *Prior to the assessment*, the data were compared to see if there were any differences in variance between the two groups. The results are shown in **Table 15.9.**

Table 15.9. Client Attendance in Fifteen Similar Programs.

Supervisor A	Supervisor B
93	79
83	82
97	73
102	85
141	143
95	86
77	85
97	81
115	78
127	85
88	74
68	78
101	82
82	85
104	78

1. Was there any difference in the pre-test variances of the two groups in **Table 15.9**? Use a one-way ANOVA with $\alpha=.05$ to analyze the data. On the basis of the results, could the assessment continue? Explain your answer.

2. Analyze the data in **Table 15.10** with one-way ANOVA. If there is statistical significance with $\alpha = .05$, use Duncan"s Multiple Range Analysis ($\alpha=.05$) to identify where the means are significantly different.

Table 15.10. Fictitious Scores.

Treatments				
Level 1	Level 2	Level 3	Level 4	Level 5
4	5	5	8	5
5	7	4	4	4
6	9	6	6	3
7	8	5	8	4
4	9	5	5	6
6	7	6	6	4
5	10	4	7	5

3. **Table 15.11** on the next page displays data obtained from subjects randomally assigned to 4 treatment levels. Analyze the data with one-way ANOVA, and determine the means that are significant through Duncan's Multiple Range analysis. Use an α level of .05 where needed.

Table 15.11. Fictitious Data.

Treatments			
Level 1	Level 2	Level 3	Level 4
13	15	20	3
16	12	13	12
7	6	11	9
10	10	12	8
14	13	8	2
13	12	9	3
8	6	16	6
15	12	20	9
17	13	9	5
18	16	18	5

4.   The following scores were rearranged into a randomized block design and listed as **Table 15.12.** Analyze the data using a randomized block ANOVA and find out which means are different with Duncan's Multiple Range analysis. ($\alpha=.05$).

Table 15.12. A Randomized Block of Fictitious Data.

Block	Treatments				
	Level 1	Level 2	Level 3	Level 4	Level 5
1	7	8	5	8	4
2	6	7	6	6	4
3	6	9	6	6	3
4	5	10	4	7	5
5	5	7	4	4	4

5.   In 1993, a small hotel kept it's head count each month. The manager was not satisfied with the number of clients, so she decided to enlist the aid of a commercial marketing firm. The firm began their effort in 1994, and the hotel manager analyzed the results with a one-way ANOVA. Assuming ANOVA was a proper procedure, was the marketing effort successful? **Table 15.13** displays the data.

Table 15.13. Head Counts in 1993 and 1994, in Hundreds.

Month	1993	1994
Jan	9	21
Feb	10	19
Mar	18	20
Apr	14	13
Jun	18	15
Jul	5	20
Aug	8	22
Sep	11	25
Oct	12	17
Nov	13	10

Chapter 16

Multiple Analysis of Variance

1. Two Way ANOVA

hen using several different levels of treatment (differing numbers of columns or variables) while also considering amounts, kinds or levels of a different variable (rows), a randomized factorial design is appropriate. This complex method allows a number of levels of different variables to be examined at the same time.

Data are randomly assigned to each of the treatments under study. A design is made with different treatments of one variable (or different variables) placed in columns. Next, the columns are subdivided into different factors or levels of a variable different from the original. Clients are next assigned randomly to the cells. The experiment is carried out and the scores recorded for each cell. Basically, Σx^2 (called sums of squares, or SS) are then found and used as in one-way ANOVA, with some minor differences.

Example. During a month-long membership campaign, 40 staff members were assigned 50 males and 50 females to attempt to recruit. Each staff person was randomly assigned one of four marketing methods to use. At the end of the month, the number of clients who joined were compared. Was there any statistical difference in the four recruiting methods between females and males? Data was recorded in Figure 16.1.

A table was constructed (Figure 16.1) with the marketing methods as columns (A levels) and the gender as rows (B levels). Sums of deviations squared (SS) will be found for the columns, rows, cells, interactions of rows with columns (A X B), total, and expected. A source table will be used to determine F values, and the analysis made as in ANOVA. Duncan's Multiple Range Analysis can be used to find significant means, and a linearity graph provides further interpretation. The problem in Figure 16.1 and it's solution follows.

264 Multiple ANOVA

	Marketing Method									
	A_1		A_2		A_3		A_4			
	X_1	X^2_1	X_2	X^2_2	X_3	X^2_3	X_4	X^2_4	X_T	X^2_T
Females	39	1,521	36	1,296	39	1,521	41	1,681		
	28	784	42	1,764	40	1,600	39	1,521		
B_1	35	1,225	39	1,521	43	1,849	40	1,600		
	33	1,089	32	1,024	39	1,521	43	1,849		
	33	1,089	34	1,156	35	1,225	41	1,681		
Σ	168	5,708	183	6,761	196	7,716	204	8,332	751	28,517
Males	42	1,764	47	2,209	48	2,304	48	2,304		
	36	1,296	39	1,521	40	1,600	39	1,521		
B_2	34	1,156	38	1,444	41	1,681	42	1,764		
	31	961	33	1,089	33	1,089	37	1,369		
	30	900	37	1,369	34	1,156	37	1,369		
Σ	173	6,077	194	7,632	196	7,830	203	8,327	766	29,866
$\Sigma\Sigma$	341	11,785	377	14,393	392	15,546	407	16,659	1,517	58,383

Figure 16.1. Marketing Data.

1. $SS_A = \dfrac{341^2}{10} + \dfrac{377^2}{10} + \dfrac{392^2}{10} + \dfrac{407^2}{10} - \dfrac{1,517^2}{40}$

$SS_A = 11,628.10 + 14,211.90 + 15,366.40 + 16,564.90$

$-57,532.225$

$SS_A = 57,7772.30 - 57,532.225 = 240.075$

2. $SS_B = \dfrac{751^2}{20} + \dfrac{766^2}{20} - \dfrac{1517^2}{40} = 5.625$

3. $SS_A + SS_B = 240.075 + 5.625 = 245.70$

4. $SS_{AXB} = SS_{cells} - (SS_A + SS_B)$

$$SS_{cells} = \frac{168^2}{5} + \frac{173^2}{5} + \frac{183^2}{5} + \frac{194^2}{5} + \frac{196^2}{5} + \frac{196^2}{5}$$

$$+ \frac{204^2}{5} + \frac{203^2}{5} - \frac{1,517^2}{40}$$

$SS_{cells} = 5,644.80 + 5,985.80 + 6,697.80 + 7,527.20$

$$+7,683.20 + 8,323.20 + 8,241.80 - 57,532.2$$
$SS_{cells} = 254.775$

$SS_{AXB} = 254.775 - (240.075 + 5.625) = 254.775 - 245.700$

$SS_{AXB} = 9.075$

5. $SS_T = \Sigma X_T^2 - \frac{(\Sigma X_T)^2}{N_T} = 58,383 - \frac{1,517^2}{40} = 58,383 - 57,532.23$

$SS_T = 850.775$

6. $SS_E = SS_T - (SS_A + SS_B + SS_{AXB}) = 850.775 - (240.075 + 5.625$
$+9.075) = 596.000$

Table 16.1. Source Table for Marketing Data.

Source	df	SS	MS	F	p
A	3	240.08	80.03	4.30	>.05
B	1	5.63	5.63	0.30	n.s.
AXB	3	9.08	3.03	0.16	n.s.
E	32	596.00	18.63		
T	39	850.78	21.81		

Degrees of Freedom (df). The df are found by looking at each individual source entry in the source table. For A, there are 4 levels, so df for A are N-1 (4-1) or 3. For B, there are two levels of B, so df for B are N-1 (2-1) or 1.

A X B is a little more complicated, in that these interactions involve separate variables of A X B. In this case, df = k-1 X r-1, where k is the number of A levels (columns) and r is the number of B levels (rows). Therefore, df A X B are 3 X 1 or 3.

The next df found is total (df$_T$). Df$_T$ is the total number of scores minus 1. Since there are 40 scores, df$_T$ = 40-1 = 39.

Now df E (error or within) can be found by adding df for A (3), B (1), A x B (3) and subtracting the result (7) from df$_T$ (39). This indirect way assumes that all df added has to equal df$_T$.

F Ratio. The F value is found the same way as in one-way ANOVA. The A, B, and A x B MS's are divided by MS E (18.63) to find the appropriate F. The table is entered with the MS for A, B or A x B as the column value and df E as the row df. F is interpreted the same way as one-way ANOVA.

Discussion. The results show that in this fictitious problem, only the A level is significant. Since this level represents the different marketing methods, which one(s) are statistically different? As in one-way ANOVA, the analysis is gross; that is, it does indicate a difference between levels, but does not identify the specific level. One way to answer this question is to examine the means of each level.

Duncan's Multiple Range Analysis will be used for this purpose. The preliminary information needed follows.

1. $\overline{X}_1$ = 34.1; $\overline{X}_2$ = 37.7; $\overline{X}_3$ = 39.2; $\overline{X}_4$ = 40.7

2. $s_{\overline{x}}E = \dfrac{s(\sqrt{MS_{within}}\ or\ E\ )}{\sqrt{N}} = \dfrac{\sqrt{18.63}}{\sqrt{10}} = \dfrac{4.32}{3.16} = 1.36$

Using the information from the preceding page, a table of means is constructed below.

Table 16.2. Table of Means For Level A.

		$\bar{X_1}$	$\bar{X_2}$	$\bar{X_3}$	$\bar{X_4}$	Short. Stude. Range	Table Value
		34.1	37.7	39.2	40.7		
$\bar{X_1}$	34.1	-	3.6	5.1	6.6	3.93	R_2 2.888
$\bar{X_2}$	37.7		-	1.5	3.0	4.13	R_3 3.035
$\bar{X_3}$	39.2			-	1.5	4.26	R_4 3.131
		1	2	3	4		
	1						
	2			X	X		
	3				X		

The table of means was constructed in the manner described in Chapter 15. Using k means (3) for row 1 and 40-4 = 36 df, the table values were recorded. Since there was no table value for 36 df, the more stringent value for 30 df was used. Table values were found for the .05 level of significance.

The table values were then multiplied by the standard deviation for the standard error of the mean expected (within) to obtain the shortest studentizied range. Any mean that equals or exceeds the shortest studentizied range wa significant. **Table 16.2** shows means 2, 3 & 4 were significantly different from mean 1. No other comparisons were statistically different. The data may be interpreted as showing each of the marketing levels as different.

B Levels. In this example, the marketing effect on females and males were similar. If it had been different, the analysis could be made by inspection, because there were only two levels involved.

AXB Levels. The AXB interactions are more complex to analyze. Since these were not significant, no further analysis is necessary. The AXB mean square is describing the peculiar result of marketing methods interacting with gender. Sometimes interactions are nonsense, but at other times they do make sense. There may be times when marketing is more effective when combined with gender.

From this example and others discussed, it is apparent that ANOVA is gross, and will require further analysis in some cases to understand the results more thoroughly.

As mentioned previously, the type of organization may dictate the number of subjects required, as in the randomized block design.

There are many other ways to analyze variance. Chapter 15 used one-way ANOVA in several designs. This chapter included two levels, such as the 4 X 2 design. Expanding the 4 X 2 design to more values of levels, such as a 6 X 8, does not change the procedure except a need to add the additional column or row variances in the appropriate places. Although no examples will be given, ANOVA can be made in three levels, such as a 2 X 2 X 2, and even beyond.

Linearity. One additional way to analyze the data is to check significant levels for linearity. Since the A level was statistically different, the means are plotted on a graph to see if there is a linear trend. Since the marketing levels were designed to increase proportionately, the expectation is that the trend should be linear. However, Figure 16.2 on the following page shows the trend is not linear, due primarily to A_1.

When more than two levels of any data are statistically different, it may be well to plot the means in this manner as a linearity check.

2. Non-Parametric ANOVA

P arametric ANOVA requires the data to be at least interval or higher values. What happens when ordinal data are used? Can a statistical tool analyze variance using ordinal data? One solution lies in the Friedman two-way analysis of variance technique.

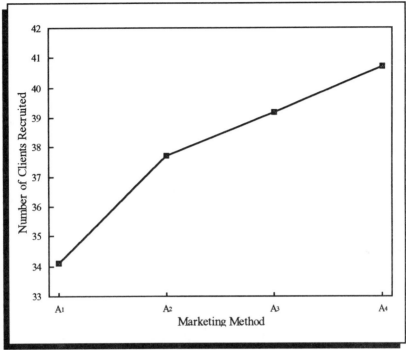

Figure 16.2. Linearity in Level A.

Using the Friedman technique, data is obtained and placed in a table, with data recorded across levels. Data could also be matched on some variable, such as age, then randomally assigned to levels or rows. The data is then ranked, with the sums of the ranks determined for each column. The column ranks are squared and used in a formula to determine chi-square. A special table is needed to test chi-square for significance.

Example: participants in a class of social recreation (N = 21) were given the Cowell Personal Distance Inventory (see Appendix A) in order to record peer rating changes during a 7 week period. Each person then led and participated in party games designed to create social interaction over 23 activity meetings of 1 hour each. The same test was given at the end of 3 weeks, and again at the end of 7 weeks. Figure 16.3 displays the scores obtained.

	Raw Score			Ranked Score		
Subject	Test 1	Test 2	Test 3	Test 1	Test 2	Test 3
1	90	87	81	3	2	1
2	88	88	70	2.5	2.5	1
3	96	98	76	2	3	1
4	102	95	72	3	2	1
5	101	101	76	2.5	2.5	1
6	82	75	51	3	2	1
7	77	83	74	2	3	1
8	86	89	76	2	3	1
9	95	92	81	3	2	1
10	87	82	77	3	2	1
11	102	93	79	3	2	1
12	97	96	91	3	2	1
13	101	91	74	3	2	1
14	84	79	68	3	2	1
15	98	96	89	3	2	1
16	97	98	78	2	3	1
17	108	95	88	3	2	1
18	76	75	65	3	2	1
19	106	101	79	3	2	1
20	97	93	73	3	2	1
21	101	103	97	2	3	1
		Σ	R_j	57	48	21
		Σ	R_j^2	3249	2304	441

Figure 16.3. Sociometric Scores of Each Student.

Procedure: first, rank the scores in each row, as in Figure 16.3. Then, solve for x^2 using the procedure found in equation 42 on the following page.

Finally, use **Table 8** in Appendix B to test for significance. Entering **Table 8** with df (n) = (k-1 = 3-1 = 2), the table value needed is 13.815 for significance at the .001 level. Since the chi-square obtained is 33.43, the data are significant beyond the .001 level. **Tables 7a** and **7b** in Appendix B are more accurate for small numbers of columns (k) and rows (N).

(42)

$$\chi^2 = \left[\frac{12}{Nk(k+1)}\right]\left[\sum_{j=1}^{k} R_j^2\right] - 3N(k+1)$$

Where χ^2 = chi square

N = number of subjects

k = number of ranked columns

$\sum_{j=1}^{k} R_j^2$ = sum of each ranked column squared

1 to k columns

$$\chi^2 = \left[\frac{12}{21(3)(3+1)}\right][3249+2304+441]-3(21)(4)$$

$$\chi^2 = \left[\frac{12}{252}\right][5994]-252$$

$$\chi^2 = 285.43-252 = 33.43$$

Discussion. The data show there is a difference between tests 1, 2, and 3. To identify where the difference lies requires logic. It is apparent the third test had the lowest rank in every case, with the second test the next lowest.

What the data seem to reveal is that as time progressed, social distance decreased. If all other factors were controlled, the class situation decreased the social distance of the students.

Also, leaders may be identified as subjects with very low scores, while isolates had very large scores. An examination of scores listed in Figure 16.3 will reveal these identities.

There are other non-parametric methods of analysis of variance also. The reader is referred to Siegel's book in the References following Appendix B for other ordinal measures.

When using data that is countable, such as sit-ups, number of baskets made, etc., other analyses can be made that may be more appropriate for the data characteristics than ANOVA, such as chi-square. Chapter 17 will show several ways to use chi-square in leisure, recreation, sport management and tourism.

APPLICATIONS

Rather than testing means, a test of variance is a more powerful tool in analyzing data. When using multiple ANOVA, data are assigned randomly. The sample size could be determined, a list of clients compiled, and the selection made randomly in assigning clientts to different treatment and/or factorial levels.

Once the experiment is concluded, analysis of the data may show gross differences, which need further interpretation. Duncan's Multiple Range Analysis or other methods, such as Scheffe', can test for significant means. Graphs of levels may check for linear trends.

ANOVA can be used for several levels in different dimensions which are not described in this text. Also, multiple designs with repeated measures are more complex and a more advanced text should be consulted for these analyses.

EXERCISES

1. Thirty two members of an age group "B" swim team were randomly assigned to either a 5,000 or 10,000 yards a day group. Each group was asked to do lat machine workouts every other day with 10 (A_1), 20 (A_2), 30 (A_3) and 40 (A_4) pounds of weight for two weeks on each trial. At the end of 4 weeks, they were timed on 25 yards freestyle. **Table 16.3** summarizes the results.

Table 16.3. Fictitious Time for 25 Yards Freestyle.

	Treatments			
	A_1	A_2	A_3	A_4
5,000 Yards	18.0	18.2	13.5	13.7
	15.0	14.3	14.3	15.0
	15.3	14.6	18.1	15.1
	14.9	12.0	20.0	18.5
10,000 Yards	16.2	18.1	17.4	12.5
	15.6	15.1	17.4	14.1
	20.0	17.6	19.2	14.9
	15.6	14.0	22.9	13.9

Using a 4 X 2 randomized factorial design, analyze the data in **Table 16.3**.

2. An experiment in which 40 clients were randomly assigned to either a 3 star or 4 star hotel while on tour was designed. In addition, they were randomly assigned to either a female or a male group leader. All clients were given a discount for agreeing to keep a log of money spent while shopping in town during the tour and participate in the study. **Table 16.4** shows the result of this fictitious experiment.

Table 16.4. Money Spent During Guided Shopping Tours.

	Type of Hotel	
	3 Star	4 Star
Female Group Leader	$180	$190
	$160	$180
	$120	$140
	$150	$140
	$140	$200
	$180	$180
	$140	$180
	$120	$180
	$170	$230
	$160	$190
Male Group Leader	$140	$210
	$120	$290
	$200	$190
	$190	$240
	$100	$210
	$130	$170
	$190	$230
	$150	$190
	$140	$190
	$100	$200

Complete the ANOVA from **Table 16.4** as a 2 X 2 randomized factorial design. Was there a statistical difference in the amount of money spent?

3. A natural site had 3 trail itineriares of equal length and difficulty. The manager wanted to know if these trails were usually traversed in the same time by different aged clients. Fifteen clients who visited the site were matched by age and randomally assigned to one of the three trails. They were asked to tour the trail, and their time was recorded in seconds. **Table 16.5** tabulates the results.

Table 16.5. Times (in Seconds) for Walking Three Trails.

Age Range	Trail 1 (A_1)	Trail 2 (A_2)	Trail 3 (A_3)
Under 24	1,646	1,533	1,598
25 - 34	1,673	1,566	1,649
35 - 44	1,922	1,781	1,890
45 - 54	1,938	1,742	1,870
55 - up	2,099	1,978	2,099

Using the Friedman Two-Way ANOVA technique, determine if the data were significantly different or not for the 3 trails. In this case, the slower time is the higher value or better rank (1 = best rank) because the manager wants to get as many groups through as quickly as possible without asking clients to hurry. *Disregarding control factors*, was there a difference in walking time for clients on the 3 trails?

Chapter 17

Chi-Square

Chi-square (χ^2) is a statistic that can be used when the data is countable. There are many examples of counting frequencies in sport management, tourism, recreation and leisure services. Examples include such things as number of patients, people with diseases, bent-knee sit-ups, wall-volley hits, people in attendance, clients served, etc. These many kinds of frequency counts lend themselves to chi-square analysis.

Assumptions. Chi-square assumes the following data characteristics:

1. countable data.
2. independence.
3. random sample.
4. continuous data.
5. distribution varies with N. As N increases, chi-square approaches normality.

Procedure. The procedure for finding chi-square is:

$$\chi^2 = \Sigma\frac{(O-E)^2}{E} = \frac{(O_1-E_1)^2}{E_1} + \frac{(O_2-E_2)^2}{E_2} + \frac{(O_3-E_3)^2}{E_3} \ldots \frac{(O_k-E_k)^2}{E_k}$$

Where: O = *Observed frequency (frequency count)*

E = *expected frequency*

k = *Number of columns (categories, factors, treatments)*

df = $k-1$ *(number of columns minus 1)*

(43)

The observed frequencies are easy to count, as they are the data gathered from observation of the criterion under study. Expected frequencies may be determined through logic or through cell probabilities.

1. One-Sample Case

Trail users were asked to register for each trail use. **Table 17.1** recorded the registrations in one month. Was there a statistical difference in trail use?

Table 17.1. Fictitious Number of Trail Users for One Month.

Trail	1	2	3	4	5	6	7	8	Total
Number Used	29	19	18	25	17	10	15	11	144

Procedure. Determine the expected frequencies (E) for each trail. Since there are eight trails, and assuming there are no differences in any trail (which there are), there should be an equal number of users on each trail. Therefore, the expected frequencies should be the total number of users (144) divided by 8 trails, or 18. Therefore, E for all eight trails is 18. Then, using equation 43:

$$\chi^2 = \sum \frac{(O-E)^2}{E} = \frac{(29-18)^2}{18} + \frac{(19-18)^2}{18} + \frac{(18-18)^2}{18} + \frac{(25-18)^2}{18}$$

$$+ \frac{(17-18)^2}{18} + \frac{(10-18)^2}{18} + \frac{(15-18)^2}{18} + \frac{(11-18)^2}{18}$$

$$\chi^2 = \frac{121}{18} + \frac{1}{18} + \frac{0}{18} + \frac{49}{18} + \frac{1}{18} + \frac{64}{18} + \frac{9}{18} + \frac{49}{18}$$

$$\chi^2 = \frac{294}{18} = 16.333$$

To determine the significance level, with k-1 degrees of freedom, df = 8-1 = 7. Entering **Table 8** (chi-square) in Appendix B with 7df (n), it was found that a table value of 14.067 is needed for significance at the .05 level. For the .02 level, a table value of 16.622 is needed. Since the χ^2 obtained is between 14.067 and 16.622, the data is significantly different between the .05 and .02 levels.

As in ANOVA, chi-square is gross. There is no way to know which lanes are significantly different. Since the average number of clients are 18, it looks like trails 1 and 4 have more clients, with trails 6 and 8 the fewer clients. Based on this fictional example, trails 1 and 4 would be the most preferred trails to hike.

2. Two Sample Case

I maginary example: numbers of clients who visited three attractions were recorded for one month to see if there were any differences between residents and non-residents who used the facilities. **Table 17.2** illustrates the *number* of users to each attraction.

Table 17.2. Number of Clients Recorded.

	Residents		Non-Residents		
Attraction	O	E	O	E	Row Σ
Museum	12	19.9	32	24.1	44
Zoo	22	16.3	14	19.7	36
Library	9	6.8	6	8.2	15
Σ	43		52		95

To determine expected frequencies (E) for each cell, first find the margin totals. They are found by adding the row and column observed (O) frequencies. Then each cell frequency expected (E) is obtained by multiplying the row and column marginal totals common to a cell and dividing by N (N = total number of cases). For example, to find E for the first place *short* cell, find the *margin totals peculiar to that cell* (row = 44, column = 43). Notice N = 95. The expected frequency is found by 44*43÷95 = 19.9. The *other expected frequencies* are found in the same manner. Then, using equation 43:

$$\chi^2 = \Sigma\frac{(O-E)^2}{E} = \frac{(12-19.9)^2}{19.9} + \frac{(32-24.1)^2}{24.1} + \frac{(22-16.3)^2}{16.3}$$

$$+\frac{(14-19.7)^2}{19.7} + \frac{(9-6.8)^2}{6.8} + \frac{(6-8.2)^2}{8.2}$$

$$\chi^2 = \frac{(7.9)^2}{19.9} + \frac{(7.9)^2}{24.1} + \frac{(5.7)^2}{16.3} + \frac{(5.7)^2}{19.7} + \frac{(2.2)^2}{6.8} + \frac{(2.2)^2}{8.2}$$

$$\chi^2 = 3.14 + 2.59 + 1.99 + 1.65 + 0.71 + 0.59 = 10.67$$

To determine the significance level **Table 17.2** reveals there are two columns and three rows. When there is more than one level, df (n) is found by (r-1)(k-1) where r = number of rows, k = number of columns. Therefore, df = (3-1)(2-1) = (2)(1) = 2.

Entering **Table 8** (chi-square) in the Appendix with 2 df (n), a table value of 9.210 is needed for significance at the .01 level. Since the chi-square obtained (10.67) is greater than the table value listed, p > .01.

Discussion. Again, the chi-square figure is gross. By inspection, it appears this unreal example favors non-residents using the museum the most, since there were 32 non-residents and only 12 residents who visited. Overall, there were 52 non-resident visits recorded, while only 43 total resident visits were registered.

3. 2 x 2 Contingency Table

**W**hen data can be placed in a 2 x 2 table, chi-square can be used to test whether or not the obtained breakdown could have occurred by chance.

Example: Eighty clients in a management workshop were asked whether or not they liked a training film. **Table 17.3** on the next page shows the results.

Table 17.3. Number of Clients Who Liked or Disliked a Training Film.

	Liked	Disliked	Σ Row
Male	(A) 15	(B) 25	40
Female	(C) 30	(D) 10	40
Σ	45	35	80

In order to find chi-square in a 2 X 2 design, a different procedure is used. The method is:

$$\chi^2 = \frac{N\left[\mid BC-AD \mid -\frac{N}{2}\right]^2}{(A+B)(C+D)(A+C)(B+D)}$$

(44)   Where: N = total frequency

BC = cell B*cell C; AD = cell A*cell D

A+B = cell A+cell B; etc.

Then, using equation 44 to find chi-square:

$$\chi^2 = \frac{80\left[\mid (25)(30)-(15)(10) \mid -\frac{80}{2}\right]^2}{(40)(40)(45)(35)} = \frac{80[\mid 750-150 \mid -40]^2}{2,520,000}$$

$$\chi^2 = \frac{80[\mid 600-40 \mid]^2}{2,520,000} = \frac{80[560]^2}{2,520,000} = \frac{25,088,000}{2,520,000} = 9.56$$

With (n) df = (k-1)(r-1) = (2-1)(2-1) = 1, the table value needed for significance at the .01 level = 6.635. Since the obtained is 9.56, p > .01. In the example above, chi-square is corrected for continuity. *The expression N÷2 (80 ÷ 2)) is the correction for continuity.*

Discussion. If this data were true, 45 persons liked the film, while 35 did not. The data indicate that the film was liked more than it was disliked, and is statistically significant (p > .01). A further analysis by inspection indicates that it was the females who liked the film the most, because 30 of 40 females liked it, whereas only 15 of 40 males liked it.

APPLICATIONS

There are numerous examples in sport management, recreation and tourism when data is recorded as frequency counts. In these cases, use of the chi-square technique is a great way to analyze the data. When a 2 by 2 contingency table can be constructed, equation 44 should be employed. Chi-square should be used only for countable observations, ie, *not actual values* of scores. The following generalizations apply: [1]

- when N > 40, use χ^2 corrected for continuity.
- when N > 20 < 40, the formula without the correction for continuity may be used if all expected frequencies > 5.
- if N < 20, the formula cannot be used. Use the Fisher Exact Test instead.

When k is larger than 2: use $(O - E)^2 \div E$ provided:

- fewer than 20% of all cells have an E < 5.
- no cell has E < 1.

---

[1] Siegel, Sidney. (1956). <u>Nonparametric statistics for the behavioral sciences,</u> (pp. 94-104). New York: Mc-Graw-Hill Book Company.

EXERCISES

1. A corporate recreation director wants to know if employees have a preference for any of 6 health films on AIDS. The following preferences were expressed by 95 employees who viewed the film. Was there a statistical difference?

Table 17.4. Film Preferences.

	Film					
	1	2	3	4	5	6
Preferences	10	18	14	19	22	12

2. The following data were obtained from a fanciful survey. Was there a difference in use between the different age groups? Explain.

Table 17.5. Use of Recreation Facilities in January.

Age	Never	Once or Twice	3-6 Times	7-12 Times	More Than 12 Times
8 - 10	5	7	6	4	4
11 - 12	8	6	9	12	11
13 - 15	12	8	6	8	7
16 - 18	14	9	5	7	4

3. A health spa manager wishes to know if female or male clients show a statistical preference for either of two treadmills. **Table 17.6** on the following page shows the number of clients using the machine in one week. Was there a statistical difference? Explain.

Table 17.6. Number of Clients Using Two Treadmills.

	Clients	
	Female	Male
Treadmill 1	40	30
Treadmill 2	24	46

4. A recreation manager wanted to know if there was a statistical difference between the number of visitors to a site from in-state and out-of-state clients who received no follow-up or a mail follow-up after the call. **Table 17.7** records the calls made during one week. Was there a statistical difference? Explain.

Table 17.7. Number of Visitors to a Site.

Type of Follow-Up	Type of Phone Call Received	
	In-State Calls	Out-Of-State-Calls
None	45	39
Mail	27	41

Appendix A

Cowell Personal Distance Inventory

PERSONAL DISTANCE INVENTORY

Personal distance is defined as how closely a person is willing to accept another person as a brother or sister. Please check the column that describes your willingness to accept the class member below as a brother or sister. All names will be held confidential by the instructor.

	I would be willing to accept her/him:						
Name	Into my family as a brother or sister 1	As a very close 'pal' or friend 2	As a member of my fraternity or sorority 3	On my street as a next door neighbor 4	Into my class at school 5	Into my school 6	Into my city 7
A	2	6	9	16	10	18	21
B	2	8	9	24	15	12	
C	2	5	3	12	30	12	7
D	1	20		17	20	6	14
E	1	12	6	16	15	12	14
F	3	18	9	16	5		
G	2	10	9	12	30	6	7
H	1	8	6	16	30	6	7
I	1	6	9	16	30	12	7
J	3	6	12	4	30	12	7
K		8	9	24	25	6	7
L		8		28	20	12	28
M	2	4	9	24	25		7
N	2	12	6	20	15	6	7
O		4	12	20	20	12	21
P	1	12	6	4	35	6	14
Q		8	6	16	25	12	21
R	3	12	3	24	10	6	7
S	1	10	9	16	15		20
T	2	12	3	12	25	12	7
U	1	4	3	10	25	6	42

[1]Modified with kind permission from Cowell, Charles C. (1958). Validating an index of social adjustment for high school use, The Research Quarterly, 29, 7-18.

Appendix B

Statistical Tables

Table 1. Table of Random Numbers[1]

Row	00000 01234	00000 56789	11111 01234	11111 56789	22222 01234	22222 56789	33333 01234	33333 56789
				First Thousand				
00	23157	54859	01837	25993	76249	70886	95230	36744
01	05545	55043	10537	43508	90611	83744	10962	21343
02	14871	60350	32404	36223	50051	00322	11543	80834
03	38976	74951	94051	75853	78805	90194	32428	71695
04	97312	61718	99755	30870	94251	25841	54882	10513
05	11742	69381	44339	30872	32797	33118	22647	06850
06	43361	28859	11016	45623	93009	00499	43640	74036
07	93806	20478	38268	04491	55751	18932	58475	52571
08	49540	13181	08429	84187	69538	29661	77738	09527
09	36768	72633	37948	21569	41959	68670	45274	83880
10	07092	52392	24627	12067	06558	45344	67338	45320
11	43310	01081	44863	80307	52555	16148	89742	94647
12	61570	06360	06173	63775	63148	95123	35017	46993
13	31352	83799	10779	18941	31579	76448	62584	86919
14	57048	86526	27795	93692	90529	56546	35065	32254
15	09243	44200	68721	07137	30729	75756	09298	27650
16	97957	35018	40894	88329	52230	82521	22532	61587
17	93732	59570	43781	98885	56671	66826	95996	44569
18	72621	11225	00922	68264	35666	59434	71687	58167
19	61020	74418	45371	20794	95917	37866	99536	19378
20	97839	85474	33055	91718	45473	54144	22034	23000
21	89160	97192	22232	90637	35055	45489	88438	16361
22	25966	88220	62871	79265	02823	52862	84919	54883
23	81443	31719	05049	54806	74690	07567	65017	16543
24	11322	54931	42362	34386	08624	97687	46245	23245

[1]Table 1 is reprinted with kind permission from Kendall, M.G., & Smith, B. Babington. (1938). Randomness and random numbers, Journal of the Royal Statistical Society,.101, 147-166.

Table 1. Table of Random Numbers[1]

Row	Column Number							
	00000 01234	00000 56789	11111 01234	11111 56789	22222 01234	22222 56789	33333 01234	33333 56789
	Second Thousand							
00	64755	83885	84122	25920	17696	15655	95045	95947
01	10302	52289	77436	34430	38112	49067	07348	23328
02	71017	98495	51308	50374	66591	02887	53765	69149
03	60012	55605	88410	34879	79655	90169	78800	03666
04	37330	94656	49161	42802	48274	54755	44553	65090
05	47869	87001	31591	12273	60626	12822	34691	61212
06	38040	42737	64167	89578	39323	49324	88434	38706
07	73508	30908	83054	80078	86669	30295	56460	45336
08	32623	46474	84061	04324	20628	37319	32356	43969
09	97591	99549	36630	35106	62069	92975	95320	57734
10	74012	31955	59790	96982	66224	24015	96749	07589
11	56754	26457	13351	05014	90966	33674	69096	33488
12	49800	49908	54831	21998	08528	26372	92923	65026
13	43584	89647	24878	56670	00221	50193	99591	62377
14	16653	79664	60325	71301	35742	83636	73058	87229
15	48502	69055	65322	58748	31446	80237	31252	96367
16	96765	54692	36316	86230	48296	38352	23816	64094
17	38923	61550	80357	81784	23444	12463	33992	28128
18	77958	81694	25225	05587	51073	01070	60218	61961
19	17928	28065	25586	08771	02641	85064	65796	48170
20	94036	85978	02318	04499	41054	10531	87431	21596
21	47460	60479	56230	48417	14372	85167	27558	00368
22	47856	56088	51992	82439	40644	17170	13463	18288
23	57616	34653	92298	62018	10375	76515	62986	90756
24	08300	92704	66752	66610	57188	79107	54222	22013

[1]Table 1 is reprinted with kind permission from Kendall, M.G., & Smith, B. Babington. (1938). Randomness and random numbers, Journal of the Royal Statistical Society. 101, 147-166.

Table 1. Table of Random Numbers[1]

Column Number							
Row 00000 01234	00000 56789	11111 01234	11111 56789	22222 01234	22222 56789	33333 01234	33333 56789

Third Thousand

Row								
00	89221	02362	65787	74733	51272	30213	92441	39651
01	04005	99818	63918	29032	94012	42363	01261	10650
02	98546	38066	50856	75045	40645	22841	53254	44125
03	41719	84401	59226	01314	54581	40398	49988	65579
04	28733	72489	00785	25843	24613	49797	85567	84471
05	65213	83927	77762	03086	80742	24395	68476	83792
06	65553	12678	90906	90466	43670	26217	69900	31205
07	05668	69080	73029	85746	58332	78231	45986	92998
08	39302	99718	49757	79515	27387	76373	47262	91612
09	64592	32254	45879	29431	38320	05981	18067	87137
10	07513	48792	47314	83660	68907	05336	82579	91582
11	86593	68501	56638	99800	82839	35148	56541	07232
12	83735	22599	97977	81248	36838	99560	32410	67614
13	08595	21826	54655	08204	87990	17033	56258	05384
14	41273	27149	44293	69458	16828	63962	15864	35431
15	00473	75908	56238	12242	72631	76314	47252	06347
16	86131	53789	81383	07868	89132	96182	07009	86432
17	33849	78359	08402	03586	03176	88663	08018	22546
18	61870	41657	07468	08612	98083	97349	20775	45091
19	43898	65923	25078	86129	78491	97653	91500	80786
20	29939	39123	04548	45985	60952	06641	28726	46473
21	38505	85555	14388	55077	18657	94887	67831	70819
22	31824	38431	67125	25511	72044	11562	53279	82268
23	91430	03767	13561	15597	06750	92552	02391	38753
24	38635	68976	25498	97526	96458	03805	04116	63514

[1]Table 1 is reprinted with kind permission from Kendall, M.G., & Smith, B. Babington. (1938). Randomness and random numbers, Journal of the Royal Statistical Society. 101, 147-166.

Table 1. Table of Random Numbers[1]

Row	00000 01234	00000 56789	11111 01234	11111 56789	22222 01234	22222 56789	33333 01234	33333 56789
				Column Number				
				Fourth Thousand				
00	02490	54122	27944	39364	94239	72074	11679	54082
01	11967	36469	60627	83701	09253	30208	01385	37482
02	48256	83465	49699	24079	05403	35154	39613	03136
03	27246	73080	21481	23536	04881	89977	49484	93071
04	32532	77265	72430	70722	86529	18457	92657	10011
05	66757	98955	92375	93431	43204	55825	45443	69265
06	11266	34545	76505	97746	34668	26999	26742	97516
07	17872	39142	45561	80146	93137	48924	64257	59284
08	62561	30365	03408	14754	51798	08133	61010	97730
09	62796	30779	35497	70501	30105	18233	00997	91970
10	75510	21771	04339	33660	42757	62223	87565	48468
11	87439	01691	63517	26590	44437	07217	98706	39032
12	97742	02621	10748	78803	38337	65226	92149	59051
13	98811	06001	21571	02875	21828	83912	85188	61624
14	58264	01852	64607	92553	29004	26695	78583	62998
15	40239	93376	10419	68610	49120	02941	80035	99317
16	26936	59186	51667	27645	46329	44681	94190	66647
17	88502	11716	98299	40974	42394	62200	69094	81646
18	63499	38093	25593	61995	79867	80569	01023	38374
19	36379	81206	03317	78710	73828	31083	60509	44091
20	93801	22322	47479	57017	59334	30647	43061	26660
21	29856	87120	56311	50053	25365	81265	22414	02431
22	97720	87931	88265	13050	71017	15177	06957	92919
23	82537	09105	74601	46377	59938	15647	34177	92753
24	75746	75268	31727	95773	72364	87324	36879	06082

[1]Table 1 is reprinted with kind permission from Kendall, M.G., & Smith, B. Babington. (1938). Randomness and random numbers, Journal of the Royal Statistical Society. 101, 147-166.

Table 1. Table of Random Numbers[1]

Row	00000 01234	00000 56789	11111 01234	11111 56789	22222 01234	22222 56789	33333 01234	33333 56789
				Column Number				
				Fifth Thousand				
00	29935	06971	63175	52579	10478	89379	61428	21363
01	15114	07126	51890	77787	75510	13103	42942	48111
02	03870	43225	10589	87629	22039	94124	38127	65022
03	79390	39188	40756	45269	65959	20640	14284	22960
04	30035	06915	79196	54428	64819	52314	48721	81594
05	29039	99861	28759	79802	68531	39198	38137	24373
06	78196	08108	24107	49777	09599	43569	84820	94956
07	15847	85493	91442	91351	80130	73752	21539	10986
08	36614	62248	49194	97209	92587	92053	41021	80064
09	40549	54884	91465	43862	35541	44466	88894	74180
10	40878	08997	14286	09982	90308	78007	51587	16658
11	10229	49282	41173	31468	59455	18756	08908	06660
12	15918	76787	30624	25928	44124	25088	31137	71614
13	13403	18796	49909	94404	64979	41462	18155	98335
14	66523	94596	74908	90271	10009	98648	17640	68909
15	91665	36469	68343	17870	25975	04662	21272	50620
16	67415	87515	08207	73729	73201	57593	96917	69699
17	76527	96996	23724	33448	63392	32394	60887	90617
18	19815	47789	74348	17147	10954	34355	81194	54407
19	25592	53587	76384	72575	84347	68918	05739	57222
20	55902	45539	63646	31609	95999	82887	40666	66692
21	02470	58376	79794	22482	42423	96162	47491	17264
22	18630	53263	13319	97619	35859	12350	14632	87659
23	89673	38230	16063	92007	59503	38402	76450	33333
24	62986	67364	06595	17427	84623	14565	82860	57300

[1]Table 1 is reprinted with kind permission from Kendall, M.G., & Smith, B. Babington. (1938). Randomness and random numbers, _Journal of the Royal Statistical Society_. _101_, 147-166.

Table 2. The Correlation Coefficient[1]

Values of the Correlation Coefficient
for Different Levels of Significance

n	.1	.05	.02	.01	.001
1	.98769	.99692	.999507	.999877	.9999988
2	.90000	.95000	.98000	.990000	.99900
3	.8054	.8783	.93433	.95873	.99116
4	.7293	.8114	.8822	.91720	.97406
5	.6694	.7545	.8329	.8745	.95074
6	.6215	.7067	.7887	.8343	.92493
7	..5822	.6664	.7498	.7977	.8982
8	.5494	.6319	.7155	.7646	.8721
9	.5214	.6021	.6851	.7348	.8471
10	.4973	.5760	.6581	.7079	.8233
11	.4762	.5529	.6339	.6835	.8010
12	.4575	.5324	.6120	.6614	.7800
13	.4409	.5139	.5923	.6411	.7603
14	.4259	.4973	.5742	.6226	.7420
15	.4124	4821	..5577	.6055	.7246
16	.4000	.4683	.5425	.5897	.77084
17	.3887	.4555	.5285	.5751	.6932
18	.3783	.4438	.5155	.5614	.6787
19	.3687	.4329	.5034	.5487	.6652
20	.3598	.4227	.4921	.5368	.6524
25	.3233	.3809	.4451	.4869	.5974
30	.2960	.3494	.4093	.4487	.5541
35	.2746	.3246	.3810	.4182	.5189
40	.2573	.3044	.3578	.3932	.4896
45	.2428	.2875	.3384	.3721	.4648
50	.2306	.2732	.3218	.3541	.4433
60	.2108	.2500	.2948	.3248	.4078
70	.1954	.2319	.2737	.3017	.3799
80	.1829	.2172	.2565	.2830	.3568
90	.1726	.2050	.2422	.2673	.3375
100	.1638	.1946	.2301	.2540	.3211

[1]Tables are taken from Tables III, IV. VII of Fisher & Yates; STATISTICAL TABLES FOR BIOLOGICAL, AGRICULTURAL AND MEDICAL RESEARCH Published by Longman Group Ltd., 1974.

Table 3. Table of Critical Values of rs,
The Spearman Rank Correlation Coefficient[1]

	Significance Level (one-tailed test)	
n	.05	.01
4	1.000	
5	.900	1.000
6	.829	.943
7	.714	.893
8	.643	.833
9	.600	.783
10	.564	.746
12	.506	.712
14	.456	.645
16	.425	.601
18	.399	.564
20	.377	.534
22	.359	.508
24	.343	.485
26	.329	.465
28	.317	.448
30	.306	.432

[1]Adapted from Olds, E.G. (1938). Distributions of sums of squares of rank differences for small numbers of individuals, Annals of Mathematical Statistics. 9, 133-148 and from Olds, E.G. (1949). The 5% significance levels for sums of squares of rank differences and a correction, Annals of Mathematical Statistics. 20, 117-118, with the kind permission of the publisher.

Table 4. Distribution of t[1]

n	.9	.8	.7	.6	.5	.4	.3	.2	.1	.05	.02	.01	.001
						Probability							
1	.158	.325	.510	.727	1.000	1.376	1.963	3.078	6.314	12.706	31.821	63.657	636.619
2	.142	.289	.445	.617	.816	1.061	1.386	1.886	2.920	4.303	6.965	9.925	31.598
3	.137	.277	.424	.584	.765	.978	1.250	1.638	2.353	3.182	4.541	5.841	12.924
4	.134	.271	.414	.569	.741	.941	1.190	1.533	2.132	2.776	3.747	4.604	8.610
5	.132	.267	.408	.559	.727	.920	1.156	1.476	2.015	2.571	3.365	4.032	6.869
6	.131	.265	.404	.553	.718	.906	1.134	1.440	1.943	2.447	3.143	3.707	5.959
7	.130	.263	.402	.549	.711	.896	1.119	1.415	1.895	2.365	2.998	3.499	5.408
8	.130	.262	.399	.546	.706	.889	1.108	1.397	1.860	2.306	2.896	3.355	5.041
9	.129	.261	.398	.543	.703	.883	1.100	1.383	1.833	2.262	2.821	3.250	4.781
10	.129	.261	.397	.542	.700	.879	1.093	1.372	1.812	2.228	2.764	3.169	4.587
11	.129	.260	.396	.540	.697	.876	1.088	1.363	1.796	2.201	2.718	3.106	4.437
12	.128	.259	.395	.539	.695	.873	1.083	1.356	1.782	2.179	2.681	3.055	4.318
13	.128	.259	.394	.538	.694	.870	1.079	1.350	1.771	2.160	2.650	3.012	4.221
14	.128	.258	.393	.537	.692	.868	1.076	1.345	1.761	2.145	2.624	2.977	4.140
15	.128	.258	.393	.536	.691	.866	1.074	1.341	1.753	2.131	2.602	2.947	4.073
16	.128	.258	.392	.535	.690	.865	1.071	1.337	1.746	2.120	2.583	2.921	4.015
17	.128	.257	.392	.534	.689	.863	1.069	1.333	1.740	2.110	2.567	2.898	3.965
18	.127	.257	.392	.534	.688	.862	1.067	1.330	1.734	2.101	2.552	2.878	3.922
19	.127	.257	.391	.533	.688	.861	1.066	1.328	1.729	2.093	2.539	2.861	3.883
20	.127	.257	.391	.533	.687	.860	1.064	1.325	1.725	2.086	2.528	2.845	3.850
21	.127	.257	.391	.532	.686	.859	1.063	1.323	1.721	2.080	2.518	2.831	3.819
22	.127	.256	.390	.532	.686	.858	1.061	1.321	1.717	2.074	2.508	2.819	3.792
23	.127	.256	.390	.532	.685	.858	1.060	1.319	1.714	2.069	2.500	2.807	3.767
24	.127	.256	.390	.531	.685	.857	1.059	1.318	1.711	2.064	2.492	2.797	3.745
25	.127	.256	.390	.531	.684	.856	1.058	1.316	1.708	2.060	2.485	2.787	3.725
26	.127	.256	.390	.531	.684	.856	1.058	1.315	1.706	2.056	2.479	2.779	3.707
27	.127	.256	.389	.531	.684	.855	1.057	1.314	1.703	2.052	2.473	2.771	3.690
28	.127	.256	.389	.530	.683	.855	1.056	1.313	1.701	2.048	2.467	2.763	3.674
29	.127	.256	.389	.530	.683	.854	1.055	1.311	1.699	2.045	2.462	2.756	3.659
30	.127	.256	.389	.530	.683	.854	1.055	1.310	1.697	2.042	2.457	2.750	3.646
40	.126	.255	.388	.529	.681	.851	1.050	1.303	1.684	2.021	2.423	2.704	3.551
60	.126	.254	.387	.527	.679	.848	1.046	1.296	1.671	2.000	2.390	2.660	3.460
120	.126	.254	.386	.526	.677	.845	1.041	1.289	1.658	1.980	2.358	2.617	3.373
∞	.126	.253	.385	.524	.674	.842	1.036	1.282	1.645	1.960	2.326	2.576	3.291

[1]Tables are taken from Tables III, IV, VII of Fisher and Yates: STATISTICAL TABLES FOR BIOLOGICAL, AGRICULTURAL AND MEDICAL RESEARCH Published by Longman Group Ltd., 1974

Table 5. Table of F. (Top row for each n = .05 level, lower row = .01 level)[1]

n_1 degrees of freedom (for greater mean square)

n	1	2	3	4	5	6	7	8	9	10	11	12	14	16	20	24	30	40	50	75	100	200	500	∞
1	161	200	216	225	230	234	237	239	241	242	243	244	245	246	248	249	250	251	252	253	254	254	254	254
	4,052	4,999	5,403	5,625	5,764	5,859	5,928	5,981	6,022	6,056	6,082	6,106	6,142	6,169	6,208	6,234	6,258	6,286	6,302	6,323	6,334	6,352	6,361	6,366
2	18.51	19.00	19.16	19.25	19.30	19.33	19.36	19.37	19.38	19.39	19.40	19.41	19.42	19.43	19.44	19.45	19.46	19.47	19.47	19.48	19.49	19.49	19.50	19.50
	98.49	99.00	99.17	99.25	99.30	99.33	99.34	99.36	99.38	99.40	99.41	99.42	99.43	99.44	99.45	99.46	99.47	99.48	99.48	99.49	99.49	99.49	99.50	99.50
3	10.13	9.55	9.28	9.12	9.01	8.94	8.88	8.84	8.81	8.78	8.76	8.74	8.71	8.69	8.66	8.64	8.62	8.60	8.58	8.57	8.56	8.54	8.54	8.53
	34.12	30.82	29.46	28.71	28.24	27.91	27.67	27.49	27.34	27.23	27.13	27.05	26.92	26.83	26.69	26.60	26.50	26.41	26.35	26.27	26.23	26.18	26.14	26.12
4	7.71	6.94	6.59	6.39	6.26	6.16	6.09	6.04	6.00	5.96	5.93	5.91	5.87	5.84	5.80	5.77	5.74	5.71	5.70	5.68	5.66	5.65	5.64	5.63
	21.20	18.00	16.69	15.98	15.52	15.21	14.98	14.80	14.66	14.54	14.45	14.37	14.24	14.15	14.02	13.93	13.83	13.74	13.69	13.61	13.57	13.52	13.48	13.46
5	6.61	5.79	5.41	5.19	5.05	4.95	4.88	4.82	4.78	4.74	4.70	4.68	4.64	4.60	4.56	4.53	4.50	4.46	4.44	4.42	4.40	4.38	4.37	4.36
	16.26	13.27	12.06	11.39	10.97	10.67	10.45	10.27	10.15	10.05	9.96	9.89	9.77	9.68	9.55	9.47	9.38	9.29	9.24	9.17	9.13	9.07	9.04	9.02
6	5.99	5.14	4.76	4.53	4.39	4.28	4.21	4.15	4.10	4.06	4.03	4.00	3.96	3.92	3.87	3.84	3.81	3.77	3.75	3.72	3.71	3.69	3.68	3.67
	13.74	10.92	9.78	9.15	8.75	8.47	8.26	8.10	7.98	7.87	7.79	7.72	7.60	7.52	7.39	7.31	7.23	7.14	7.09	7.02	6.99	6.94	6.90	6.88
7	5.59	4.74	4.35	4.12	3.97	3.87	3.79	3.73	3.68	3.63	3.60	3.57	3.52	3.49	3.44	3.41	3.38	3.34	3.32	3.29	3.28	3.25	3.24	3.23
	12.25	9.55	8.45	7.85	7.46	7.19	7.00	6.84	6.71	6.62	6.54	6.47	6.35	6.27	6.15	6.07	5.98	5.90	5.85	5.78	5.75	5.70	5.67	5.65
8	5.32	4.46	4.07	3.84	3.69	3.58	3.50	3.44	3.39	3.34	3.31	3.28	3.23	3.20	3.15	3.12	3.08	3.05	3.03	3.00	2.98	2.96	2.94	2.93
	11.26	8.65	7.59	7.01	6.63	6.37	6.19	6.03	5.91	5.82	5.74	5.67	5.56	5.48	5.36	5.28	5.20	5.11	5.06	5.00	4.96	4.91	4.88	4.86
9	5.12	4.26	3.86	3.63	3.48	3.37	3.29	3.23	3.18	3.13	3.10	3.07	3.02	2.98	2.93	2.90	2.86	2.82	2.80	2.77	2.76	2.73	2.72	2.71
	10.56	8.02	6.99	6.42	6.06	5.80	5.62	5.47	5.35	5.26	5.18	5.11	5.00	4.92	4.80	4.73	4.64	4.56	4.51	4.45	4.41	4.36	4.33	4.31
10	4.96	4.10	3.71	3.48	3.33	3.22	3.14	3.07	3.02	2.97	2.94	2.91	2.86	2.82	2.77	2.74	2.70	2.67	2.64	2.61	2.59	2.56	2.55	2.54
	10.04	7.56	6.55	5.99	5.64	5.39	5.21	5.06	4.95	4.85	4.78	4.71	4.60	4.52	4.41	4.33	4.25	4.17	4.12	4.05	4.01	3.96	3.93	3.91
11	4.84	3.98	3.59	3.36	3.20	3.09	3.01	2.95	2.90	2.86	2.82	2.79	2.74	2.70	2.65	2.61	2.57	2.53	2.50	2.47	2.45	2.42	2.41	2.40
	9.65	7.20	6.22	5.67	5.32	5.07	4.88	4.74	4.63	4.54	4.46	4.40	4.29	4.21	4.10	4.02	3.94	3.86	3.80	3.74	3.70	3.66	3.62	3.60
12	4.75	3.88	3.49	3.26	3.11	3.00	2.92	2.85	2.80	2.76	2.72	2.69	2.64	2.60	2.54	2.50	2.46	2.42	2.40	2.36	2.35	2.32	2.31	2.30
	9.33	6.93	5.95	5.41	5.06	4.82	4.65	4.50	4.39	4.30	4.22	4.16	4.05	3.98	3.86	3.78	3.70	3.61	3.56	3.49	3.46	3.41	3.38	3.36
13	4.67	3.80	3.41	3.18	3.02	2.92	2.84	2.77	2.72	2.67	2.63	2.60	2.55	2.51	2.46	2.42	2.38	2.34	2.32	2.28	2.26	2.24	2.22	2.21
	9.07	6.70	5.74	5.20	4.86	4.62	4.41	4.30	4.19	4.10	4.02	3.96	3.85	3.78	3.67	3.59	3.51	3.42	3.37	3.30	3.27	3.21	3.18	3.16

[1]Reproduced with kind permission from Snedecor, G.W. (1956). Table of F, Statistical Methods, (5th Edition). Ames: Iowa State College Press.

Table 5. Table of F. (Top row for each n = .05 level, lower row = .01 level)[1]

n_1 degrees of freedom (for greater mean square)

n	1	2	3	4	5	6	7	8	9	10	11	12	14	16	20	24	30	40	50	75	100	200	500	∞
14	4.60	3.74	3.34	3.11	2.96	2.85	2.77	2.70	2.65	2.60	2.56	2.53	2.48	2.44	2.39	2.35	2.31	2.27	2.24	2.21	2.19	2.16	2.14	2.13
	8.86	6.51	5.56	5.03	4.69	4.46	4.28	4.14	4.03	3.94	3.86	3.80	3.70	3.62	3.51	3.43	3.34	3.26	3.21	3.14	3.11	3.06	3.02	3.00
15	4.54	3.68	3.29	3.06	2.90	2.79	2.70	2.64	2.59	2.55	2.51	2.48	2.43	2.39	2.33	2.29	2.25	2.21	2.18	2.15	2.12	2.10	2.08	2.07
	8.68	6.36	5.42	4.89	4.56	4.32	4.14	4.00	3.89	3.80	3.73	3.67	3.56	3.48	3.36	3.29	3.20	3.12	3.07	3.00	2.97	2.92	2.89	2.87
16	4.49	3.63	3.24	3.01	2.85	2.74	2.66	2.59	2.54	2.49	2.45	2.42	2.37	2.33	2.28	2.24	2.20	2.16	2.13	2.09	2.07	2.04	2.02	2.01
	8.53	6.23	5.29	4.77	4.44	4.20	4.03	3.89	3.78	3.69	3.61	3.55	3.45	3.37	3.25	3.18	3.10	3.01	2.96	2.89	2.86	2.80	2.77	2.75
17	4.45	3.59	3.20	2.96	2.81	2.70	2.62	2.55	2.50	2.45	2.41	2.38	2.33	2.29	2.23	2.19	2.15	2.11	2.08	2.04	2.02	1.99	1.97	1.96
	8.40	6.11	5.18	4.67	4.34	4.10	3.93	3.79	3.68	3.59	3.52	3.45	3.35	3.27	3.16	3.08	3.00	2.92	2.86	2.79	2.76	2.70	2.67	2.65
18	4.41	3.55	3.16	2.93	2.77	2.66	2.58	2.51	2.46	2.41	2.37	2.34	2.29	2.25	2.19	2.15	2.11	2.07	2.04	2.00	1.98	1.95	1.93	1.92
	8.28	6.01	5.09	4.58	4.25	4.01	3.85	3.71	3.60	3.51	3.44	3.37	3.27	3.19	3.07	3.00	2.91	2.83	2.78	2.71	2.68	2.62	2.59	2.57
19	4.38	3.52	3.13	2.90	2.74	2.63	2.55	2.48	2.43	2.38	2.34	2.31	2.26	2.21	2.15	2.11	2.07	2.02	2.00	1.96	1.94	1.91	1.90	1.88
	8.18	5.93	5.01	4.50	4.17	3.94	3.77	3.63	3.52	3.43	3.36	3.30	3.19	3.12	3.00	2.92	2.84	2.76	2.70	2.63	2.60	2.54	2.51	2.49
20	4.35	3.49	3.10	2.87	2.71	2.60	2.52	2.45	2.40	2.35	2.31	2.28	2.23	2.18	2.12	2.08	2.04	1.99	1.96	1.92	1.90	1.87	1.85	1.84
	8.10	5.85	4.94	4.43	4.10	3.87	3.71	3.56	3.45	3.37	3.30	3.23	3.13	3.05	2.94	2.86	2.77	2.69	2.63	2.56	2.53	2.47	2.44	2.42
21	4.32	3.47	3.07	2.84	2.68	2.57	2.49	2.42	2.37	2.32	2.28	2.25	2.20	2.15	2.09	2.05	2.00	1.96	1.93	1.89	1.87	1.84	1.82	1.81
	8.02	5.78	4.87	4.37	4.04	3.81	3.65	3.51	3.40	3.31	3.24	3.17	3.07	2.99	2.88	2.80	2.72	2.63	2.58	2.51	2.47	2.42	2.38	2.36
22	4.30	3.44	3.05	2.82	2.66	2.55	2.47	2.40	2.35	2.30	2.26	2.23	2.18	2.13	2.07	2.03	1.98	1.93	1.91	1.87	1.84	1.81	1.80	1.78
	7.94	5.72	4.82	4.31	3.99	3.76	3.59	3.45	3.35	3.26	3.18	3.12	3.02	2.94	2.83	2.75	2.67	2.58	2.53	2.46	2.42	2.37	2.33	2.31
23	4.28	3.42	3.03	2.80	2.64	2.53	2.45	2.38	2.32	2.28	2.24	2.20	2.14	2.10	2.04	2.00	1.96	1.91	1.88	1.84	1.82	1.79	1.77	1.76
	7.88	5.66	4.76	4.26	3.94	3.71	3.54	3.41	3.30	3.21	3.14	3.07	2.97	2.89	2.78	2.70	2.62	2.53	2.48	2.41	2.37	2.32	2.28	2.26
24	4.26	3.40	3.01	2.78	2.62	2.51	2.43	2.36	2.30	2.26	2.22	2.18	2.13	2.09	2.02	1.98	1.94	1.89	1.86	1.82	1.80	1.76	1.74	1.73
	7.82	5.61	4.72	4.22	3.90	3.67	3.50	3.36	3.25	3.17	3.09	3.03	2.93	2.85	2.74	2.66	2.58	2.49	2.44	2.36	2.33	2.27	2.23	2.21
25	4.24	3.38	2.99	2.76	2.60	2.49	2.41	2.34	2.28	2.24	2.20	2.16	2.11	2.06	2.00	1.96	1.92	1.87	1.84	1.80	1.77	1.74	1.72	1.71
	7.77	5.57	4.68	4.18	3.86	3.63	3.46	3.32	3.21	3.13	3.05	2.99	2.89	2.81	2.70	2.62	2.54	2.45	2.40	2.32	2.29	2.23	2.19	2.17
26	4.22	3.37	2.98	2.74	2.59	2.47	2.39	2.32	2.27	2.22	2.18	2.15	2.10	2.05	1.99	1.95	1.90	1.85	1.82	1.78	1.76	1.72	1.70	1.69
	7.72	5.53	4.64	4.14	3.82	3.59	3.42	3.29	3.17	3.09	3.02	2.96	2.86	2.77	2.66	2.58	2.50	2.41	2.36	2.28	2.25	2.19	2.15	2.13

[1]Reproduced with kind permission from Snedecor, G.W. (1956). Table of F, Statistical Methods, (5th Edition). Ames: Iowa State College Press.

Table 5. Table of F. (Top row for each n = .05 level, lower row = .01 level)[1]

n₁ degrees of freedom (for greater mean square)

n	1	2	3	4	5	6	7	8	9	10	11	12	14	16	20	24	30	40	50	75	100	200	500	∞
27	4.21	3.35	2.96	2.73	2.57	2.46	2.37	2.30	2.25	2.20	2.16	2.13	2.08	2.03	1.97	1.93	1.88	1.84	1.80	1.76	1.74	1.71	1.68	1.67
	7.68	5.49	4.60	4.11	3.79	3.56	3.39	3.26	3.14	3.06	2.98	2.93	2.83	2.74	2.63	2.55	2.47	2.38	2.33	2.25	2.21	2.16	2.12	2.10
28	4.20	3.31	2.95	2.71	2.56	2.44	2.36	2.29	2.24	2.19	2.15	2.12	2.06	2.02	1.96	1.91	1.87	1.81	1.78	1.75	1.72	1.69	1.67	1.65
	7.64	5.45	4.57	4.07	3.76	3.53	3.36	3.23	3.11	3.03	2.95	2.90	2.80	2.71	2.60	2.52	2.44	2.35	2.30	2.22	2.18	2.13	2.09	2.06
29	4.18	3.33	2.93	2.70	2.54	2.43	2.35	2.28	2.22	2.18	2.14	2.10	2.05	2.00	1.94	1.90	1.85	1.80	1.77	1.73	1.71	1.68	1.65	1.64
	7.60	5.42	4.54	4.04	3.73	3.50	3.33	3.20	3.08	3.00	2.92	2.87	2.77	2.68	2.57	2.49	2.41	2.32	2.27	2.19	2.15	2.10	2.06	2.03
30	4.17	3.32	2.92	2.69	2.53	2.42	2.34	2.27	2.21	2.16	2.12	2.09	2.04	1.99	1.93	1.89	1.84	1.79	1.76	1.72	1.69	1.66	1.64	1.62
	7.56	5.39	4.51	4.02	3.70	3.47	3.30	3.17	3.06	2.98	2.90	2.84	2.74	2.66	2.55	2.47	2.38	2.29	2.24	2.16	2.13	2.07	2.03	2.01
32	4.15	3.30	2.90	2.67	2.51	2.40	2.32	2.25	2.19	2.14	2.10	2.07	2.02	1.97	1.91	1.86	1.82	1.76	1.74	1.69	1.67	1.64	1.61	1.59
	7.50	5.34	4.46	3.97	3.66	3.42	3.25	3.12	3.01	2.94	2.86	2.80	2.70	2.62	2.51	2.42	2.34	2.25	2.20	2.12	2.08	2.02	1.98	1.96
34	4.13	3.28	2.88	2.65	2.49	2.38	2.30	2.23	2.17	2.12	2.08	2.05	2.00	1.95	1.89	1.84	1.80	1.74	1.71	1.67	1.64	1.61	1.59	1.57
	7.44	5.29	4.42	3.93	3.61	3.38	3.21	3.08	2.97	2.89	2.82	2.76	2.66	2.58	2.47	2.38	2.30	2.21	2.15	2.08	2.04	1.98	1.94	1.91
36	4.11	3.26	2.86	2.63	2.48	2.36	2.28	2.21	2.15	2.10	2.06	2.03	1.98	1.93	1.87	1.82	1.78	1.72	1.69	1.65	1.62	1.59	1.56	1.55
	7.39	5.25	4.38	3.89	3.58	3.35	3.18	3.04	2.94	2.86	2.78	2.72	2.62	2.54	2.43	2.35	2.26	2.17	2.12	2.04	2.00	1.94	1.90	1.87
38	4.10	3.25	2.85	2.62	2.46	2.35	2.26	2.19	2.14	2.09	2.05	2.02	1.96	1.92	1.85	1.80	1.76	1.71	1.67	1.63	1.60	1.57	1.54	1.53
	7.35	5.21	4.34	3.86	3.54	3.32	3.15	3.02	2.91	2.82	2.75	2.69	2.59	2.51	2.40	2.32	2.22	2.14	2.08	2.00	1.97	1.90	1.86	1.84
40	4.08	3.23	2.84	2.61	2.45	2.34	2.25	2.18	2.12	2.07	2.04	2.00	1.95	1.90	1.84	1.79	1.74	1.69	1.66	1.61	1.59	1.55	1.53	1.51
	7.31	5.18	4.31	3.83	3.51	3.29	3.12	2.99	2.88	2.80	2.73	2.66	2.56	2.49	2.37	2.29	2.20	2.11	2.05	1.97	1.94	1.88	1.84	1.81
42	4.07	3.22	2.83	2.59	2.44	2.32	2.24	2.17	2.11	2.06	2.02	1.99	1.94	1.89	1.82	1.78	1.73	1.68	1.64	1.60	1.57	1.54	1.51	1.49
	7.27	5.15	4.29	3.80	3.49	3.26	3.10	2.96	2.86	2.77	2.70	2.64	2.54	2.46	2.35	2.26	2.17	2.08	2.02	1.94	1.91	1.85	1.80	1.78
44	4.06	3.21	2.82	2.58	2.43	2.31	2.23	2.16	2.10	2.05	2.01	1.98	1.92	1.88	1.81	1.76	1.72	1.66	1.63	1.58	1.56	1.52	1.50	1.48
	7.24	5.12	4.26	3.78	3.46	3.24	3.07	2.94	2.84	2.75	2.68	2.62	2.52	2.44	2.32	2.24	2.15	2.06	2.00	1.92	1.88	1.82	1.78	1.75
46	4.05	3.20	2.81	2.57	2.42	2.30	2.22	2.14	2.09	2.04	2.00	1.97	1.91	1.87	1.80	1.75	1.71	1.65	1.62	1.57	1.54	1.51	1.48	1.46
	7.21	5.10	4.24	3.76	3.44	3.22	3.05	2.92	2.82	2.73	2.66	2.60	2.50	2.42	2.30	2.22	2.13	2.04	1.98	1.90	1.86	1.80	1.76	1.72
48	4.04	3.19	2.80	2.56	2.41	2.30	2.21	2.14	2.08	2.03	1.99	1.96	1.90	1.86	1.79	1.74	1.70	1.64	1.61	1.56	1.53	1.50	1.47	1.45
	7.19	5.08	4.22	3.74	3.42	3.20	3.04	2.92	2.80	2.71	2.64	2.58	2.48	2.40	2.28	2.20	2.11	2.02	1.96	1.88	1.84	1.78	1.73	1.70

[1]Reproduced with kind permission from Snedecor, G.W. (1956). Table of F, Statistical Methods, (5th Edition). Ames: Iowa State College Press.

Table 5. Table of F. (Top row for each n = .05 level, lower row = .01 level)[1]

n₁ degrees of freedom (for greater mean square)

n	1	2	3	4	5	6	7	8	9	10	11	12	14	16	20	24	30	40	50	75	100	200	500	∞
50	4.03	3.18	2.79	2.56	2.40	2.29	2.20	2.13	2.07	2.02	1.98	1.95	1.90	1.85	1.78	1.74	1.69	1.63	1.60	1.55	1.52	1.48	1.46	1.44
	7.17	5.06	4.20	3.72	3.41	3.18	3.02	2.88	2.78	2.70	2.62	2.56	2.46	2.39	2.26	2.18	2.10	2.00	1.94	1.86	1.82	1.76	1.71	1.68
55	4.02	3.17	2.78	2.54	2.38	2.27	2.18	2.11	2.05	2.00	1.97	1.93	1.88	1.83	1.76	1.72	1.67	1.61	1.58	1.52	1.50	1.46	1.43	1.41
	7.12	5.01	4.16	3.68	3.37	3.15	2.98	2.85	2.75	2.66	2.59	2.53	2.43	2.35	2.23	2.15	2.06	1.96	1.90	1.82	1.78	1.71	1.66	1.64
60	4.00	3.15	2.76	2.52	2.37	2.25	2.17	2.10	2.04	1.99	1.95	1.92	1.86	1.81	1.75	1.70	1.65	1.59	1.56	1.50	1.48	1.44	1.41	1.39
	7.08	4.98	4.13	3.65	3.34	3.12	2.95	2.82	2.72	2.63	2.56	2.50	2.40	2.32	2.20	2.12	2.03	1.93	1.87	1.79	1.74	1.68	1.63	1.60
65	3.99	3.14	2.75	2.51	2.36	2.24	2.15	2.08	2.02	1.98	1.94	1.90	1.85	1.80	1.73	1.68	1.63	1.57	1.54	1.49	1.46	1.42	1.39	1.37
	7.04	4.95	4.10	3.62	3.31	3.09	2.93	2.79	2.70	2.61	2.54	2.47	2.37	2.30	2.18	2.09	2.00	1.90	1.84	1.76	1.71	1.64	1.60	1.56
70	3.98	3.13	2.74	2.50	2.35	2.23	2.14	2.07	2.01	1.97	1.93	1.89	1.84	1.79	1.72	1.67	1.62	1.56	1.53	1.47	1.45	1.40	1.37	1.35
	7.01	4.92	4.08	3.60	3.29	3.07	2.91	2.77	2.67	2.59	2.51	2.45	2.35	2.28	2.15	2.07	1.98	1.88	1.82	1.74	1.69	1.62	1.56	1.53
80	3.96	3.11	2.72	2.48	2.33	2.21	2.12	2.05	1.99	1.95	1.91	1.88	1.82	1.77	1.70	1.65	1.60	1.54	1.51	1.45	1.42	1.38	1.35	1.32
	6.96	4.88	4.04	3.56	3.25	3.04	2.87	2.74	2.64	2.55	2.48	2.41	2.32	2.24	2.11	2.03	1.94	1.84	1.78	1.70	1.65	1.57	1.52	1.49
100	3.94	3.09	2.70	2.46	2.30	2.19	2.10	2.03	1.97	1.92	1.88	1.85	1.79	1.75	1.68	1.63	1.57	1.51	1.48	1.42	1.39	1.34	1.30	1.28
	6.90	4.82	3.98	3.51	3.20	2.99	2.82	2.69	2.59	2.51	2.43	2.36	2.26	2.19	2.06	1.98	1.89	1.79	1.73	1.64	1.59	1.51	1.46	1.43
125	3.92	3.07	2.68	2.44	2.29	2.17	2.08	2.01	1.95	1.90	1.86	1.83	1.77	1.72	1.65	1.60	1.55	1.49	1.45	1.39	1.36	1.31	1.27	1.25
	6.84	4.78	3.94	3.47	3.17	2.95	2.79	2.65	2.56	2.47	2.40	2.33	2.23	2.15	2.03	1.94	1.85	1.75	1.68	1.59	1.54	1.46	1.40	1.37
150	3.91	3.06	2.67	2.43	2.27	2.16	2.07	2.00	1.94	1.89	1.85	1.82	1.76	1.71	1.64	1.59	1.54	1.47	1.44	1.37	1.34	1.29	1.25	1.22
	6.81	4.75	3.91	3.44	3.14	2.92	2.76	2.62	2.53	2.44	2.37	2.30	2.20	2.12	2.00	1.91	1.83	1.72	1.66	1.56	1.51	1.43	1.37	1.33
200	3.89	3.04	2.65	2.41	2.26	2.14	2.05	1.98	1.92	1.87	1.83	1.80	1.74	1.69	1.62	1.57	1.52	1.45	1.42	1.35	1.32	1.26	1.22	1.19
	6.76	4.71	3.88	3.41	3.11	2.90	2.73	2.60	2.50	2.41	2.34	2.28	2.17	2.09	1.97	1.88	1.79	1.69	1.62	1.53	1.48	1.39	1.33	1.28
400	3.86	3.02	2.62	2.39	2.23	2.12	2.03	1.96	1.90	1.85	1.81	1.78	1.72	1.67	1.60	1.54	1.49	1.42	1.38	1.32	1.28	1.22	1.16	1.13
	6.70	4.66	3.83	3.36	3.06	2.85	2.69	2.55	2.46	2.37	2.29	2.23	2.12	2.04	1.92	1.84	1.74	1.64	1.57	1.47	1.42	1.32	1.24	1.19
1000	3.85	3.00	2.61	2.38	2.22	2.10	2.02	1.95	1.89	1.84	1.80	1.76	1.70	1.65	1.58	1.53	1.47	1.41	1.36	1.30	1.26	1.19	1.13	1.08
	6.66	4.62	3.80	3.34	3.04	2.82	2.66	2.53	2.43	2.34	2.26	2.20	2.09	2.01	1.89	1.81	1.71	1.61	1.54	1.44	1.38	1.28	1.19	1.11
∞	3.84	2.99	2.60	2.37	2.21	2.09	2.01	1.94	1.88	1.83	1.79	1.75	1.69	1.64	1.57	1.52	1.46	1.40	1.35	1.28	1.24	1.17	1.11	1.00
	6.64	4.60	3.78	3.32	3.02	2.80	2.64	2.51	2.41	2.32	2.24	2.18	2.07	1.99	1.87	1.79	1.69	1.59	1.52	1.41	1.36	1.25	1.15	1.00

[1]Reproduced with kind permission from Snedecor, G.W. (1956). Table of F, Statistical Methods, (5th Edition). Ames: Iowa State College Press.

Table 6a. Significant Studentized Ranges for Duncan's New Multiple Text with $\alpha = .10$[1]

df\k	2	3	4	5	6	7	8	9	10	11	12	13	14	15	16	17	18	19
2	4.130																	
3	3.328	3.330																
4	3.015	3.074	3.081															
5	2.850	2.934	2.964	2.970														
6	2.748	2.846	2.890	2.908	2.911													
7	2.680	2.785	2.838	2.864	2.876	2.878												
8	2.630	2.742	2.800	2.832	2.849	2.857	2.858											
9	2.592	2.708	2.771	2.808	2.829	2.840	2.845	2.847										
10	2.563	2.682	2.748	2.788	2.813	2.827	2.835	2.839	2.839									
11	2.540	2.660	2.730	2.772	2.799	2.817	2.827	2.833	2.835	2.835								
12	2.521	2.643	2.714	2.759	2.789	2.808	2.821	2.828	2.832	2.833	2.833							
13	2.505	2.628	2.701	2.748	2.779	2.800	2.815	2.824	2.829	2.832	2.832	2.832						
14	2.491	2.616	2.690	2.739	2.771	2.794	2.810	2.820	2.827	2.831	2.832	2.833	2.833					
15	2.479	2.605	2.681	2.731	2.765	2.789	2.805	2.817	2.825	2.830	2.833	2.834	2.834	2.834				
16	2.469	2.596	2.673	2.723	2.759	2.784	2.802	2.815	2.824	2.829	2.833	2.835	2.836	2.836	2.836			
17	2.460	2.588	2.665	2.717	2.753	2.780	2.798	2.812	2.822	2.829	2.834	2.836	2.838	2.838	2.838	2.838		
18	2.452	2.580	2.659	2.712	2.749	2.776	2.796	2.810	2.821	2.828	2.834	2.838	2.840	2.840	2.840	2.840	2.840	
19	2.445	2.574	2.653	2.707	2.745	2.773	2.793	2.808	2.820	2.828	2.834	2.839	2.841	2.842	2.843	2.843	2.843	2.843
20	2.439	2.568	2.648	2.702	2.741	2.770	2.791	2.807	2.819	2.828	2.834	2.839	2.843	2.845	2.845	2.845	2.845	2.845
24	2.420	2.550	2.632	2.688	2.729	2.760	2.783	2.801	2.816	2.827	2.835	2.842	2.848	2.851	2.854	2.856	2.857	2.857
30	2.400	2.532	2.615	2.674	2.717	2.750	2.776	2.796	2.813	2.826	2.837	2.846	2.853	2.859	2.863	2.867	2.869	2.871
40	2.831	2.514	2.600	2.660	2.705	2.741	2.769	2.791	2.810	2.825	2.838	2.849	2.858	2.866	2.873	2.878	2.883	2.887
60	2.363	2.497	2.584	2.646	2.694	2.731	2.761	2.786	2.807	2.825	2.839	2.853	2.864	2.874	2.883	2.890	2.897	2.903
120	2.344	2.479	2.568	2.632	2.682	2.722	2.754	2.781	2.804	2.824	2.842	2.857	2.871	2.883	2.893	2.903	2.912	2.920
∞	2.326	2.462	2.552	2.619	2.670	2.712	2.746	2.776	2.801	2.824	2.844	2.861	2.877	2.892	2.905	2.918	2.929	2.939

[1]Reproduced from: H.L. Harter, "Critical Values for Duncan's New Multiple Range Test." BIOMETRICS 16: 671-685, 1960. With kind permission from the Biometric Society.

Table 6b. Significant Studentized Ranges for Duncan's New Multiple Text with α = .05[1]

df \ k	2	3	4	5	6	7	8	9	10	11	12	13	14	15	16	17	18	19
2	6.085																	
3	4.501	4.516																
4	3.927	4.013	4.033															
5	3.635	3.749	3.797	3.814														
6	3.461	3.587	3.649	3.680	3.694													
7	3.344	3.477	3.548	3.588	3.611	3.622												
8	3.261	3.399	3.475	3.521	3.549	3.566	3.575											
9	3.199	3.339	3.420	3.470	3.502	3.523	3.536	3.544										
10	3.151	3.293	3.376	3.430	3.465	3.489	3.505	3.516	3.522									
11	3.113	3.256	3.342	3.397	3.435	3.462	3.480	3.493	3.501	3.506								
12	3.082	3.225	3.313	3.370	3.410	3.439	3.459	3.474	3.484	3.491	3.496							
13	3.055	3.200	3.289	3.348	3.389	3.419	3.442	3.458	3.470	3.478	3.484	3.488						
14	3.033	3.178	3.268	3.329	3.372	3.403	3.426	3.444	3.457	3.467	3.474	3.479	3.482					
15	3.014	3.160	3.250	3.312	3.356	3.389	3.413	3.432	3.446	3.457	3.465	3.471	3.476	3.478				
16	2.998	3.144	3.235	3.298	3.343	3.376	3.402	3.422	3.437	3.449	3.458	3.465	3.470	3.473	3.477			
17	2.984	3.130	3.222	3.285	3.331	3.366	3.392	3.412	3.429	3.441	3.451	3.459	3.465	3.469	3.473	3.475		
18	2.971	3.118	3.210	3.274	3.321	3.356	3.383	3.405	3.421	3.435	3.445	3.454	3.460	3.465	3.470	3.472	3.474	
19	2.960	3.107	3.199	3.264	3.311	3.347	3.375	3.397	3.415	3.429	3.440	3.449	3.456	3.462	3.467	3.470	3.472	3.473
20	2.950	3.097	3.190	3.255	3.303	3.339	3.368	3.391	3.409	3.424	3.436	3.445	3.453	3.459	3.464	3.467	3.470	3.472
24	2.919	3.066	3.160	3.226	3.276	3.315	3.345	3.370	3.390	3.406	3.420	3.432	3.441	3.449	3.456	3.461	3.465	3.469
30	2.888	3.035	3.131	3.199	3.250	3.290	3.322	3.349	3.371	3.389	3.405	3.418	3.430	3.439	3.447	3.454	3.460	3.466
40	2.858	3.006	3.102	3.171	3.224	3.266	3.300	3.328	3.352	3.373	3.390	3.405	3.418	3.429	3.439	3.448	3.456	3.463
60	2.829	2.976	3.073	3.143	3.198	3.241	3.277	3.307	3.333	3.355	3.374	3.391	3.406	3.419	3.431	3.442	3.451	3.460
120	2.800	2.947	3.045	3.116	3.172	3.217	3.254	3.287	3.314	3.337	3.359	3.377	3.394	3.409	3.423	3.435	3.446	3.457
∞	2.772	2.918	3.017	3.089	3.146	3.193	3.232	3.265	3.294	3.320	3.343	3.363	3.382	3.399	3.414	3.428	3.442	3.454

[1] Reproduced from: H.L. Harter, "Critical Values for Duncan's New Multiple Range Test." BIOMETRICS 16: 671-685. 1960. With kind permission from the Biometric Society.

Table 6c. Significant Studentized Ranges for Duncan's New Multiple Text with $\alpha = .01$[1]

df \ k	2	3	4	5	6	7	8	9	10	11	12	13	14	15	16	17	18	19
2	14.04																	
3	8.261	8.321																
4	6.512	6.677	6.740															
5	5.702	5.893	5.989	6.040														
6	5.243	5.439	5.549	5.614	5.655													
7	4.949	5.145	5.260	5.334	5.383	5.416												
8	4.746	4.939	5.057	5.135	5.189	5.227	5.256											
9	4.596	4.787	4.906	4.986	5.043	5.086	5.118	5.142										
10	4.482	4.671	4.790	4.871	4.931	4.975	5.010	5.037	5.058									
11	4.392	4.579	4.697	4.780	4.841	4.887	4.924	4.952	4.975	4.994								
12	4.320	4.504	4.622	4.706	4.767	4.815	4.852	4.883	4.907	4.927	4.944							
13	4.260	4.442	4.560	4.644	4.706	4.755	4.793	4.824	4.850	4.872	4.889	4.904						
14	4.210	4.391	4.508	4.591	4.654	4.704	4.743	4.775	4.802	4.824	4.843	4.859	4.872					
15	4.168	4.347	4.463	4.547	4.610	4.660	4.700	4.733	4.760	4.783	4.803	4.820	4.834	4.846				
16	4.131	4.309	4.425	4.509	4.572	4.622	4.663	4.696	4.724	4.748	4.768	4.786	4.800	4.813	4.825			
17	4.099	4.275	4.391	4.475	4.539	4.589	4.630	4.664	4.693	4.717	4.738	4.756	4.771	4.785	4.797	4.807		
18	4.071	4.246	4.362	4.445	4.509	4.560	4.601	4.635	4.664	4.689	4.711	4.729	4.745	4.759	4.772	4.783	4.792	
19	4.046	4.220	4.335	4.419	4.483	4.534	4.575	4.610	4.639	4.665	4.686	4.705	4.722	4.736	4.749	4.761	4.771	4.780
20	4.024	4.197	4.312	4.395	4.459	4.510	4.552	4.587	4.617	4.642	4.664	4.684	4.701	4.716	4.729	4.741	4.751	4.761
24	3.956	4.126	4.239	4.322	4.386	4.437	4.480	4.516	4.546	4.573	4.596	4.616	4.634	4.651	4.665	4.678	4.690	4.700
30	3.889	4.056	4.168	4.250	4.314	4.366	4.409	4.445	4.477	4.504	4.528	4.550	4.569	4.586	4.601	4.615	4.628	4.640
40	3.825	3.988	4.098	4.180	4.244	4.296	4.339	4.376	4.408	4.436	4.461	4.483	4.503	4.521	4.537	4.553	4.566	4.579
60	3.762	3.922	4.031	4.111	4.174	4.226	4.270	4.307	4.340	4.368	4.394	4.417	4.438	4.456	4.474	4.490	4.504	4.518
120	3.702	3.858	3.965	4.044	4.107	4.158	4.202	4.239	4.272	4.301	4.327	4.351	4.372	4.392	4.410	4.426	4.442	4.456
∞	3.643	3.796	3.900	3.978	4.040	4.091	4.135	4.172	4.205	4.235	4.261	4.285	4.307	4.327	4.345	4.363	4.379	4.394

[1] Reproduced from: H.L. Harter, "Critical Values for Duncan's New Multiple Range Test." BIOMETRICS 16: 671-685. 1960. With kind permission from the Biometric Society.

Table 6d. Significant Studentized Ranges for Duncan's New Multiple Text with α = .005[1]

df \ k	2	3	4	5	6	7	8	9	10	11	12	13	14	15	16	17	18	19
2	19.93																	
3	10.55	10.63																
4	7.916	8.126	8.210															
5	6.751	6.980	7.100	7.167														
6	6.105	6.334	6.466	6.547	6.600													
7	5.699	5.922	6.057	6.145	6.207	6.250												
8	5.420	5.638	5.773	5.864	5.930	5.978	6.014											
9	5.218	5.430	5.565	5.657	5.725	5.776	5.815	5.846										
10	5.065	5.273	5.405	5.498	5.567	5.620	5.662	5.695	5.722									
11	4.945	5.149	5.280	5.372	5.442	5.496	5.539	5.574	5.603	5.626								
12	4.849	5.048	5.178	5.270	5.341	5.396	5.439	5.475	5.505	5.531	5.552							
13	4.770	4.966	5.094	5.186	5.256	5.312	5.356	5.393	5.424	5.450	5.472	5.492						
14	4.704	4.897	5.023	5.116	5.185	5.241	5.286	5.324	5.355	5.382	5.405	5.425	5.442					
15	4.647	4.838	4.964	5.055	5.125	5.181	5.226	5.264	5.297	5.324	5.348	5.368	5.386	5.402				
16	4.599	4.787	4.912	5.003	5.073	5.129	5.175	5.213	5.245	5.273	5.298	5.319	5.338	5.354	5.368			
17	4.557	4.744	4.867	4.958	5.027	5.084	5.130	5.168	5.201	5.229	5.254	5.275	5.295	5.311	5.327	5.340		
18	4.521	4.705	4.828	4.918	4.987	5.043	5.090	5.129	5.162	5.190	5.215	5.237	5.256	5.274	5.289	5.303	5.316	
19	4.489	4.671	4.793	4.883	4.952	5.008	5.054	5.093	5.127	5.156	5.181	5.203	5.222	5.240	5.256	5.270	5.283	5.295
20	4.460	4.641	4.762	4.851	4.920	4.976	5.022	5.061	5.095	5.124	5.150	5.172	5.193	5.210	5.226	5.241	5.254	5.266
24	4.371	4.547	4.666	4.753	4.822	4.877	4.924	4.963	4.997	5.027	5.053	5.076	5.097	5.116	5.133	5.148	5.162	5.175
30	4.285	4.456	4.572	4.658	4.726	4.781	4.827	4.867	4.901	4.931	4.958	4.981	5.003	5.022	5.040	5.056	5.071	5.085
40	4.202	4.369	4.482	4.566	4.632	4.687	4.733	4.772	4.806	4.837	4.864	4.888	4.910	4.930	4.948	4.965	4.980	4.995
60	4.122	4.284	4.394	4.476	4.541	4.595	4.640	4.679	4.713	4.744	4.771	4.796	4.818	4.838	4.857	4.874	4.890	4.905
120	4.045	4.201	4.308	4.388	4.452	4.505	4.550	4.588	4.622	4.652	4.679	4.704	4.726	4.747	4.766	4.784	4.800	4.815
∞	3.970	4.121	4.225	4.303	4.365	4.417	4.461	4.499	4.532	4.562	4.589	4.614	4.636	4.657	4.676	4.694	4.710	4.726

[1]Reproduced from: H.L. Harter, "Critical Values for Duncan's New Multiple Range Test." BIOMETRICS 16: 671-685. 1960. With kind permission from the Biometric Society.

Table 6e. Significant Studentized Ranges for Duncan's New Multiple Text with α = .001[1]

df \ k	2	3	4	5	6	7	8	9	10	11	12	13	14	15	16	17	18	19
2	44.69																	
3	18.28	18.45																
4	12.18	12.52	12.67															
5	9.714	10.05	10.24	10.35														
6	8.427	8.743	8.932	9.055	9.139													
7	7.648	7.943	8.127	8.252	8.342	8.409												
8	7.130	7.407	7.584	7.708	7.799	7.869	7.924											
9	6.762	7.024	7.195	7.316	7.407	7.478	7.535	7.582										
10	6.487	6.738	6.902	7.021	7.111	7.182	7.240	7.287	7.327									
11	6.275	6.516	6.676	6.791	6.880	6.950	7.008	7.056	7.097	7.132								
12	6.106	6.340	6.494	6.607	6.695	6.765	6.822	6.870	6.911	6.947	6.978							
13	5.970	6.195	6.346	6.457	6.543	6.612	6.670	6.718	6.759	6.795	6.826	6.854						
14	5.856	6.075	6.223	6.332	6.416	6.485	6.542	6.590	6.631	6.667	6.699	6.727	6.752					
15	5.760	5.974	6.119	6.225	6.309	6.377	6.433	6.481	6.522	6.558	6.590	6.619	6.644	6.666				
16	5.678	5.888	6.030	6.135	6.217	6.284	6.340	6.388	6.429	6.465	6.497	6.525	6.551	6.574	6.595			
17	5.608	5.813	5.953	6.056	6.138	6.204	6.260	6.307	6.348	6.384	6.416	6.444	6.470	6.493	6.514	6.533		
18	5.546	5.748	5.886	5.988	6.068	6.134	6.189	6.236	6.277	6.313	6.345	6.373	6.399	6.422	6.443	6.462	6.480	
19	5.492	5.691	5.826	5.927	6.007	6.072	6.127	6.174	6.214	6.250	6.281	6.310	6.336	6.359	6.380	6.400	6.418	6.434
20	5.444	5.640	5.774	5.873	5.952	6.017	6.071	6.117	6.158	6.193	6.225	6.254	6.279	6.303	6.324	6.344	6.362	6.379
24	5.297	5.484	5.612	5.708	5.784	5.846	5.899	5.945	5.984	6.020	6.051	6.079	6.105	6.129	6.150	6.170	6.188	6.205
30	5.156	5.335	5.457	5.549	5.622	5.682	5.734	5.778	5.817	5.851	5.882	5.910	5.935	5.958	5.980	6.000	6.018	6.036
40	5.022	5.191	5.308	5.396	5.466	5.524	5.574	5.617	5.654	5.688	5.718	5.745	5.770	5.793	5.814	5.834	5.852	5.869
60	4.894	5.055	5.166	5.249	5.317	5.372	5.420	5.461	5.498	5.530	5.559	5.586	5.610	5.632	5.653	5.672	5.690	5.707
120	4.771	4.924	5.029	5.109	5.173	5.226	5.271	5.311	5.346	5.377	5.405	5.431	5.454	5.476	5.496	5.515	5.532	5.549
∞	4.654	4.798	4.898	4.974	5.034	5.085	5.128	5.166	5.199	5.229	5.256	5.280	5.303	5.324	5.343	5.361	5.378	5.394

[1]Reproduced from: H.L. Harter, "Critical Values for Duncan's New Multiple Range Test." BIOMETRICS 16: 671–685. 1960. With kind permission from the Biometric Society.

Table 7a. Table of Probabilities Associated with Values as Large as Observed Values of χ_r^2 in the Friedman Two-Way Analysis of Variance by Ranks[1]

Table 7a k = 3

N = 2		N = 3		N = 4		N = 5	
χ_r^2	p	χ_r^2	p	χ_r^2	p	χ_r^2	p
0	1.000	.000	1.000	.0	1.000	.0	1.000
1	.833	.667	.944	.5	.931	.4	.954
3	.500	2.000	.528	1.5	.653	1.2	.691
4	.167	2.667	.361	2.0	.431	1.6	.522
		4.667	.194	3.5	.273	2.8	.367
		6.000	.028	4.5	.125	3.6	.182
				6.0	.069	4.8	.124
				6.5	.042	5.2	.093
				8.0	.0046	6.4	.039
						7.6	.024
						8.4	.0085
						10.0	.00077

N = 6		N = 7		N = 8		N = 9	
χ_r^2	p	χ_r^2	p	χ_r^2	p	χ_r^2	p
.00	1.000	.000	1.000	.00	1.000	.000	1.000
.33	.956	.286	.964	.25	.967	.222	.971
1.00	.740	.857	.768	.75	.794	.667	.814
1.33	.570	1.143	.620	1.00	.654	.889	.865
2.33	.430	2.000	.486	1.75	.531	1.556	.569
3.00	.252	2.571	.305	2.25	.355	2.000	.398
4.00	.184	3.429	.237	3.00	.285	2.667	.328
4.33	.142	3.714	.192	3.25	.236	2.889	.278
5.33	.072	4.571	.112	4.00	.149	3.556	.187
6.33	.052	5.429	.085	4.75	.120	4.222	.154
7.00	.029	6.000	.052	5.25	.079	4.667	.107
8.33	.012	7.143	.027	6.25	.047	5.556	.069
9.00	.0081	7.714	.021	6.75	.038	6.000	.057
9.33	.0055	8.000	.016	7.00	.030	6.222	.048
10.33	.0017	8.857	.0084	7.75	.018	6.889	.031
12.00	.00013	10.286	.0036	9.00	.0099	8.000	.019
		10.571	.0027	9.25	.0080	8.222	.016
		11.143	.0012	9.75	.0048	8.667	.010
		12.286	.00032	10.75	.0024	9.556	.0060
		14.000	.000021	12.00	.0011	10.667	.0035
				12.25	.00086	10.889	.0029
				13.00	.00026	11.556	.0013
				14.25	.000061	12.667	.00066
				16.00	.0000036	13.556	.00035
						14.000	.00020
						14.222	.000097
						14.889	.000054
						16.222	.000011
						18.000	.0000006

Table 7b. Table of Probabilities Associated with Values as Large as Observed Values of χ_r^2 in the Friedman Two-Way Analysis of Variance by Ranks[1]

Table 7b k = 4

N = 2		N = 3		N = 4			
χ_r^2	p	χ_r^2	p	χ_r^2	p	χ_r^2	p
.0	1.000	.2	1.000	.0	1.000	5.7	.141
.6	.958	.6	.958	.3	.992	6.0	.105
1.2	.834	1.0	.910	.6	.928	6.3	.094
1.8	.792	1.8	.727	.9	.900	6.6	.077
2.4	.625	2.2	.608	1.2	.800	6.9	.068
3.0	.542	2.6	.524	1.5	.754	7.2	.054
3.6	.458	3.4	.446	1.8	.677	7.5	.052
4.2	.375	3.8	.342	2.1	.649	7.8	.036
4.8	.208	4.2	.300	2.4	.524	8.1	.033
5.4	.167	5.0	.207	2.7	.508	8.4	.019
6.0	.042	5.4	.175	3.0	.432	8.7	.014
		5.8	.148	3.3	.389	9.3	.012
		6.6	.075	3.6	.355	9.6	.0069
		7.0	.054	3.9	.324	9.9	.0062
		7.4	.033	4.5	.242	10.2	.0027
		8.2	.017	4.8	.200	10.8	.0016
		9.0	.0017	5.1	.190	11.1	.00094
				5.4	.158	12.0	.000072

[1]Reproduced and adapted from: Friedman, M. (1937). The use of ranks to avoid the assumption of normality implicit in the analysis of variance, _Journal American Statistical Association_. 32, 675-701. with kind permission from the American Statistical Association.

Table 8. Distribution of χ^2
Probability.

n	.99	.98	.95	.90	.80	.70	.50	.30	.20	.10	.05	.02	.01	.001
1	$.0^3157$	$.0^3628$	.00393	.0158	.0642	.148	.455	1.074	1.642	2.706	3.841	5.412	6.635	10.827
2	.0201	.0404	.103	.211	.446	.713	1.386	2.408	3.219	4.605	5.991	7.824	9.210	13.815
3	.115	.185	.352	.584	1.005	1.424	2.366	3.665	4.642	6.251	7.815	9.837	11.345	16.266
4	.297	.429	.711	1.064	1.649	2.195	3.357	4.878	5.989	7.779	9.488	11.668	13.277	18.467
5	.554	.752	1.145	1.610	2.343	3.000	4.351	6.064	7.289	9.236	11.070	13.388	15.086	20.515
6	.872	1.134	1.635	2.204	3.070	3.828	5.348	7.231	8.558	10.645	12.592	15.033	16.812	22.457
7	1.239	1.564	2.167	2.833	3.822	4.671	6.346	8.383	9.803	12.017	14.067	16.622	18.475	24.322
8	1.646	2.032	2.733	3.490	4.594	5.527	7.344	9.524	11.030	13.362	15.507	18.168	20.090	26.125
9	2.088	2.532	3.325	4.168	5.380	6.393	8.343	10.656	12.242	14.684	16.919	19.679	21.666	27.877
10	2.558	3.059	3.940	4.865	6.179	7.267	9.342	11.781	13.442	15.987	18.307	21.161	23.209	29.588
11	3.053	3.609	4.575	5.578	6.989	8.148	10.341	12.899	14.631	17.275	19.675	22.618	24.725	31.264
12	3.571	4.178	5.226	6.304	7.807	9.034	11.340	14.011	15.812	18.549	21.026	24.054	26.217	32.909
13	4.107	4.765	5.892	7.042	8.634	9.926	12.340	15.119	16.985	19.812	22.362	25.472	27.688	34.528
14	4.660	5.368	6.571	7.790	9.467	10.821	13.339	16.222	18.151	21.064	23.685	26.873	29.141	36.123
15	5.229	5.985	7.261	8.547	10.307	11.721	14.339	17.322	19.311	22.307	24.996	28.259	30.578	37.697
16	5.812	6.614	7.962	9.312	11.152	12.624	15.338	18.418	20.465	23.542	26.296	29.633	32.000	39.252
17	6.408	7.255	8.672	10.085	12.002	13.531	16.338	19.511	21.615	24.769	27.587	30.995	33.409	40.790
18	7.015	7.906	9.390	10.865	12.857	14.440	17.338	20.601	22.760	25.989	28.869	32.346	34.805	42.312
19	7.633	8.567	10.117	11.651	13.716	15.352	18.338	21.689	23.900	27.204	30.144	33.687	36.191	43.820
20	8.260	9.237	10.851	12.443	14.578	16.266	19.337	22.775	25.038	28.412	31.410	35.020	37.366	45.315
21	8.897	9.915	11.591	13.240	15.445	17.182	20.337	23.858	26.171	29.615	32.671	36.343	38.932	46.797
22	9.542	10.600	12.338	14.041	16.314	18.101	21.337	24.939	27.301	30.813	33.924	37.659	40.289	48.268
23	10.196	11.293	13.091	14.848	17.187	19.021	22.337	26.018	28.429	32.007	35.172	38.968	41.638	49.728
24	10.856	11.992	13.848	15.659	18.062	19.943	23.337	27.096	29.553	33.196	36.415	40.270	42.980	51.179
25	11.524	12.697	14.611	16.473	18.940	20.867	24.337	28.172	30.675	34.382	37.652	41.566	44.314	52.620
26	12.198	13.409	15.379	17.292	19.820	21.792	25.336	29.246	31.795	35.563	38.885	42.856	45.642	54.052
27	12.879	14.125	16.151	18.114	20.703	22.719	26.336	30.319	32.912	36.471	40.113	44.140	46.963	55.476
28	13.565	14.847	16.928	18.939	21.588	23.647	27.336	31.391	34.027	37.916	41.337	45.419	48.278	56.893
29	14.256	15.574	17.708	19.768	22.475	24.577	28.336	32.461	35.139	39.087	42.557	46.693	49.588	58.302
30	14.953	16.306	18.493	20.599	23.364	25.508	29.336	33.530	36.250	40.256	43.773	47.962	50.892	59.703
32	16.362	17.783	20.072	22.271	25.148	27.373	31.336	35.665	38.466	42.585	46.194	50.487	53.486	62.487
34	17.789	19.275	21.664	23.952	26.938	29.242	33.336	37.795	40.676	44.903	48.602	52.995	56.061	65.247
36	19.233	20.783	23.269	25.643	28.735	31.115	35.336	39.922	42.879	47.212	50.999	55.489	58.619	67.985
38	20.691	22.304	24.884	27.343	30.537	32.992	37.335	42.045	45.076	49.513	53.384	57.696	61.162	70.703
40	22.164	23.838	26.509	29.051	32.345	34.872	39.335	44.165	47.269	51.805	55.759	60.436	63.691	73.402
42	23.650	25.383	28.144	30.765	34.157	36.755	41.335	46.282	49.456	54.090	58.124	62.892	66.206	76.084
44	25.148	26.939	29.787	32.487	35.974	38.641	43.335	48.396	51.639	56.369	60.481	65.337	68.710	78.750
46	26.657	28.504	31.439	34.215	37.795	40.529	45.335	50.507	53.818	58.641	62.830	67.771	71.201	81.400
48	28.177	30.080	33.098	35.949	39.621	42.420	47.335	52.616	55.993	60.907	65.171	70.197	73.683	84.037
50	29.707	31.664	34.764	37.689	41.449	44.313	49.335	54.723	58.164	63.167	67.505	72.613	76.154	86.661
52	31.246	33.256	36.437	39.433	43.281	46.209	51.335	56.827	60.332	65.422	69.832	75.021	78.616	89.272
54	32.793	34.856	38.116	41.183	45.117	48.106	53.335	58.930	62.496	67.673	72.153	77.422	81.069	91.872
56	34.350	36.464	39.801	42.937	46.955	50.005	55.335	61.031	64.658	69.919	74.468	79.815	83.513	94.461
58	35.913	38.078	41.492	44.696	48.797	51.906	57.335	63.129	66.816	72.160	76.778	82.201	85.950	97.039
60	37.485	39.699	43.188	46.459	50.641	53.809	59.335	65.227	68.972	74.397	79.082	84.580	88.379	99.607
62	39.063	41.327	44.889	48.226	52.487	55.714	61.335	67.322	71.125	76.630	81.381	86.953	90.802	102.166
64	40.649	42.960	46.595	49.996	54.336	57.620	63.335	69.416	73.276	78.860	83.675	89.320	93.217	104.716
66	42.240	44.599	48.305	51.770	56.188	59.527	65.335	71.508	75.424	81.085	85.965	91.681	95.626	107.258
68	43.838	46.244	50.020	53.548	58.042	61.436	67.335	73.600	77.571	83.308	88.250	94.037	98.028	109.791
70	45.442	47.893	51.739	55.329	59.898	63.346	69.334	75.689	79.715	85.527	90.531	96.388	100.425	112.317

For odd values of n between 30 and 70 the mean of the tabular values for $n-1$ and $n+1$ may be taken. For larger values of n. the expression $\sqrt{2\chi^2}-\sqrt{2n-1}$ may be used as a normal deviate with unit variance, remembering that the probability for χ^2 corresponds with that of a single tail of the normal curve. (For fuller formulae see Introduction.) Tables are taken from Tables III, IV, VII of Fisher & Yates: STATISTICAL TABLES FOR BIOLOGICAL, AGRICULTURAL AND MEDICAL RESEARCH published by Longman Group UK Ltd., 1974.

References

American Psychological Association. (1985). Publication manual of the American Psychological Association (3rd ed.). Washington, DC: American Psychological Association.

Bethel, John P. (Gen. ed.). (1949). Webster's new collegiate dictionary (p. 270). Springfield, MA: G. & C. Merriam Company.

Bookwalter, Karl W. (1965). Selection and definition of a problem. In M. Gladys Scott (Ed.), Research methods in health, physical education, recreation (pp. 39-63). Washington, DC: American Association for Health, Physical Education and Recreation.

Campbell, D.T. & Stanley, J.C. (1963). Experimental and quasi-experimental designs for research. Boston: Houghton Mifflin Company.

Carter, Edward. (1967, Spring Qaurter). [Class notes taken in the course History and Philosophy of Education, East Carolina University].

Colbert, C.C. (1970). A comparative study of a jogging program versus a running program for improvement of cardiovascular fitness. Unpublished master's thesis, The university of Georgia, Athens.

Corbin, Dan. (1970). Recreation leadership. Englewood Cliffs, New Jersey: Prentice-Hall, Inc.

Cowell, Charles C. (1958). Validating an index of social adjustment for high school use. The Research Quarterly, 29, 7-18.

Edwards, Allen L. (1968). (3rd. ed.). Experimental design in psychological research. New York: Holt, Rinehart and Winston, Inc.

Friedman, M. (1937). The use of ranks to avoid the assumption of normality implicit in the analysis of variance. Journal of the American Statistical Association, 32, 675-701.

Fisher, Sir Ronald A., & Yates, Frank. (1974). Statistical tables for biological agricultural and medical research. London: Longman Group Limited.

Goeldner, C.R. & Dikie, Karen. (1980). Bibliography of tourism and travel research studies, reports and articles. Boulder, CO: Colorado Business Research Division, College of Business, University of Colorado.

Jubenville, Alan. (1976). Outdoor recreation planning. Philadelphia: W.B. Saunders Company.

Kachigan, Sam K. (1986). Statistical Analysis. New York: Radius Press.

Kendall, M.G., & Smith, B. Babington. (1938). Randomness and random sampling numbers. Journal of the Royal Statistical Society, 101, 147-167.

Latham, John. (1991). Bias due to group size in surveys. Journal of travel research, XXIX (4), 32-35.

Likert, R. (1970). A technique for the measurement of attitudes. In G.F. Summers (Ed.), Attitude measurement (pp. 149-158). Chicago: Rand-McNally & Company.

Metheny, Eleanor. (1965). Philosophical methods. In M. Gladys Scott (Ed.), Research methods in health, physical education, recreation (pp. 482-496). Washington: American Association for Health, Physical Education and Recreation.

Meyers, J.L. (1969). Fundementals of experimental design. Boston: Allyn & Bacon.

Olds, E.G. (1938). Distribution of sums of squares of rank differences for a small number of individuals. Annals of Mathematical Statistics, 9, 133-148.

Olds, E.G. (1949). The 5% significance levels for sums of squares of rank differences and a correction. Annals of Mathematical Statistics, 20, 117-118.

Osgood, J.C. Suci, C. & Tannebaum, P. (1957). The measurement of meaning. Urbana: The University of Illinois Press.

Rarick, Lawrence. (1965). The case study. In M. Gladys Scott (Ed.), Research methods in health, physical education, recreation. Washington: American Association for Health, Physical Education and Recreation.

Reid, Laurel J. & Andeereck, Kathleen L. (Fall, 1989). Statistical analyses used in tourism research. Journal of Travel Research, 21-24.

Seigel, Sidney. (1956). Nonparametric statistics for the behavorial sciences. New York: McGraw-Hill Book Company.

Seashore, H.G. (ed.). (January, 1955). Test Service Bulletin (p.8). New York: The Psychological Corporation.

Shockley, Joe M. Jr. (1971). A critical analysis of the 1971 NCAA swimming championships, with a description of personal variables and training methods. Unpublished doctoral dissertation, The University of Georgia, Athens.

Shockley, Joe M. Jr. (Fall, 1977). A sociometric analysis of a recreation activity. Geogria Association for Health, Physical Education, Recreation and Dance, 13-16.

Shockley, Joe M. Jr. (Sept., 1977). Warmup with a cold shower?, Track Technique. 2195-2197.

Siegel, Sidney. (1956). Nonparametric statistics for the behavioral sciences. New York: Mcgraw-Hill Book Company.

Snedecor, G.W. (1956). Table of F. Statistical methods (5th ed.). Ames: Iowa State College Press.

Steinhaus, Arthur H. (1955). Why this research? In M. Gladys Scott (Ed.), Research methods in health, physical education, recreation. Washington: American Association for Health, Physical Education and Recreation.

Strobel, C. F. (1958). Introduction to elementary statistics. Raleigh: The Technical Press, North Carolina State University.

Thurstone, L.L. (1970). Attitudes can be measured. In G.G. Summers (Ed.), Attitude Measurement (pp. 127-141). Chicago: Rand McNally & Company.

Thomas, Jerry R. & Nelson, Jack K. (1990). Research methods in physical activity. Champaign, Illinois: Human Kinetics Books.

U.S. Bureau of the Census. (1990). Statistical abstract of the United States. Washington, DC: U.S. Bureau of the Census.

U.S. Department of Commerce. (1994). U.S. industrial outlook, 1994 . Washington, DC: U.S. Department opf Commerce.

Van Buren, Abigail. (1985, March 27). No Title. The Caledonian Record. p. 13.

Van Dalen, D.B. (1965). The historical method, In M. Gladys Scott (Ed.), Research methods in health, physical education, recreation. Washington: American Association for Health, Physical Education and Recreation.

Walsh, Robert (1991). (Pub.). Walker's manual of western publications (pp. 378, 781). San Mateo, California: Walker's Western Research.

Williams, D.C. Jr. (1985). Sample size for research in tourism. Visions in Leisure and Business, 4 (2), 19-29.

Index